Springer Biographies

The books published in the Springer Biographies tell of the life and work of scholars, innovators, and pioneers in all fields of learning and throughout the ages. Prominent scientists and philosophers will feature, but so too will lesser known personalities whose significant contributions deserve greater recognition and whose remarkable life stories will stir and motivate readers. Authored by historians and other academic writers, the volumes describe and analyse the main achievements of their subjects in manner accessible to nonspecialists, interweaving these with salient aspects of the protagonists' personal lives. Autobiographies and memoirs also fall into the scope of the series.

More information about this series at http://www.springer.com/series/13617

Shinya Inoué

Pathways of a Cell Biologist

Through Yet Another Eye

Springer

Shinya Inoué
Marine Biological Laboratory
Woods Hole, MA, USA

Springer Biographies
ISBN 978-981-10-0946-4 ISBN 978-981-10-0947-1 (eBook)
DOI 10.1007/978-981-10-0947-1

Library of Congress Control Number: 2016949742

Cover caption: Shinya investigates living cells with his polarized light microscope

Printed on acid-free paper

This Springer imprint is published by Springer Nature
The registered company is Springer Science+Business Media Singapore Pte Ltd.

Foreword

Having had the honor of receiving the 2008 Nobel Prize in Chemistry, I, Osamu Shimomura, am now being presented the additional honor to prepare a foreword for Prof. Shinya Inoué's autobiography *Pathways of a Cell Biologist.* Although having to rely on his autobiography as far as his personal life, especially as related to his youth, I am more or less familiar with Shinya's contribution to science, especially since we have been coworkers at the Marine Biological Laboratory (MBL) in Woods Hole for some 34 years.

Born in London, England, in 1921 and raised in China (Quemoy); Portland, Oregon; and Sydney, Australia, Shinya's early language was primarily English especially since his Japanese diplomatic parents had not insisted on their use of the Japanese language at home. In fact Shinya's English was so familiar to him that he won first prize in his English composition class after arriving in Sydney after Portland. And after returning to Tokyo from Sydney, he had to catch up on his Japanese language quickly enough so that he had to repeat the 5th grade in the Magome primary school rather than enter the 6th grade in order to adequately prepare for the competitive admission to high school. Fortunately he had a jovial, understanding teacher at Magome and was admitted to the 2nd city high school in Ueno, Tokyo.

Not only was Shinya's early upbringing abroad in these English-speaking countries conducive to his interest in the natural sciences, but especially his distaste for high-school-taught Japanese history must have had clear influences on his career choice. One main saving aspect of his high school experience was the extracurricular shop class offered by an interested graduate. There Shinya delighted in learning to design and make vacuum-tube radios, amplifiers, steam engines, and other devices extracurricularly.

Shinya's interest in science, especially in deciphering the mechanism of cell division in live cells, became strongly motivated in 1941, when he was introduced to Katsuma Dan (Katy, a nickname used by himself) at Musashi Koutou Gakkou (Japanese senior high school). Katy had acquired his Ph.D. 10 years previously at the University of Pennsylvania under L.V. Heilbrunn, where he met Jean Clark, also a Ph.D., who was to be his wife. Despite this international hostile background,

Shinya and Dan became lifelong friends who collaborated on their biophysical studies of living cells over many years.

Shinya and Katy found (following W. J. Schmidt in Tubingen) the special advantage that polarized light microscopy provided for studying dynamic molecular events in living cells. Early on they (S. & K.) demonstrated how the submicroscopic fibers that made up the spindle fibers (which otherwise were not visible without fixing the cells) could be visualized with the polarizing microscope. Then in 1948 Katy and Jean managed a fellowship for Shinya to enter the graduate school at Princeton, New Jersey. Once at Princeton, Shinya's mentor, Kenneth Cooper, encouraged Shinya to follow his own interest in pursuing his thesis research.

Thus Shinya first studied changes in the birefringence of the mitotic spindle associated with cell division while concurrently adding some improvements to the polarizing microscope which led to his Ph.D. thesis. Based on those studies, he proposed how the rearrangement of the spindle molecules generated the forces that pulled chromosomes apart. These studies were backed by studies on the influence of colchicine on the morphology and birefringence of metaphase-arrested *Chaetopterus* spindles as well as the effects of temperature on the morphology and birefringence of the *Chaetopterus* developing meiotic spindles.

A dramatic improvement in the sensitivity and resolving power of the polarizing microscope was introduced a few years later by Shinya and coworkers at the American Optical Company Research Group in Southbridge, Massachusetts. What was known up to then was that even using objective and condenser lenses completely free of strain birefringence, the numerical aperture of the polarizing microscope had to be quite limited by reducing the opening of the condenser aperture in order to maintain a reasonably high extinction factor. But closing down the condenser aperture not only reduced the amount of light available through the microscope but limits its resolving power drastically.

What Shinya had known before his seminar presentation at Southbridge was that the back aperture of the polarizing microscope was only uniformly and nearly completely dark between crossed polarizers when the condenser aperture was quite closed down, say to a numerical aperture (NA) of approximately 0.3. At any larger condenser NA, the back aperture between crossed polarizers no longer was very dark but became a dark center surrounded by 4 quadrants of increasing brightness away from the center. Between the quadrants were arms of a dark cross paralleling the polarizer- and analyzer-crossed orientation.

When the polarizer and analyzer were not set at exactly 90° to each other, the dark arms of the cross opened up into two Vs in opposite quadrants. The newly formed dark Vs indicated regions of the back aperture which were now extinguished by the analyzer. And it took some 8 or 9 ° of analyzer rotation for the two Vs to reach the edge of the back aperture of a 1.25 NA condenser and objective lenses. In other words, the spherical lens surfaces of the condenser and objective lenses had induced that much rotation away from the 90 ° crossed angle set by the polarizer and analyzer. No wonder the extinction factor was drastically reduced as one opened the condenser diaphragm and let higher NA beams through.

As proposed by Lem Hyde of Nikon, a half-wave birefringent plate should reverse the rotation induced by the curved surfaces of the condenser and objective lenses. Therefore why not introduce a half-wave plate between the condenser and objective lenses and let the two rotations cancel each other out? As noted by Shinya, the half-wave plate between the condenser and objective would indeed change the rotation of light at the lens surfaces according to the curvature of each point that light entered or exited the glass surface; hence rotation by the lens surfaces in the condenser and objective would complicate but not cancel each other away.

Shinya's solution was to introduce the half-wave plate at the aperture plane (where the rays were now nearly parallel) outside of the condenser or objective. This gave rise to the polarization rectifier which in fact raised the extinction factor of a pair of 1.25 NA lenses by 20-fold, incredibly several times better than the five-fold improvement required by Dr. Brian O'Brein of the University of Rochester Optics Institute in order for American Optical to support Shinya's research to improve the extinction factor of the polarizing microscope!

With the extinction factor thus raised, and especially effectively so at high NAs, the imaging and birefringence-detecting sensitivity of the polarizing microscope were dramatically improved. Not only were the weakly birefringent microtubules in the cytoplasm clearly imaged, but the detailed patterns of the weakly birefringent fibers contouring the surface of epithelial cells now stood out for the first time. At the same time, the detailed arrangement of DNA molecules in maturing spermatocytes became analyzable for the first time.

Thus Shinya was able to acquire a series of birefringent images of maturing spermatocytes of the crane fly and from their compensated images to map out the detailed arrangement of the intracellular DNA. This led to the demonstration of the double helical structure of the maturing chromosomes, which themselves were clearly delineated along the length of the maturing spermatocytes. The intricate three-dimensional arrangement of the chromosomes thus revealed with the rectified polarizing microscope required their high-resolution birefringence analytical capability, but at such dimensions that the same chromosomes could not be visualized by electron microscopy.

In addition to the barrier-breaking findings based on the use of the rectifiers in the polarizing microscope, additional discoveries were made by Shinya and colleagues by further improvement of the birefringent-detecting sensitivity of the polarizing microscope by combining it with TV imaging and electronic image processing. For these developments, the physiology course at the Marine Biological Laboratory in Woods Hole played important roles. In 1963, Shinya and Bob Allen gave their first physiology course presentations which, together with the faculty, they had assembled for those occasions. During the course demonstrations, both Bob and Shinya and their faculty found the surprising improvement that image contrast could be vastly improved by passing the image output of the microscope through a TV system.

In more recent advances, we must pay special tribute to the contribution of many specialists from optical, electronic, and computer companies and to the new physicists specializing in electronics. Their contributions made by many specialists with a variety of physical backgrounds have not only opened the eyes of other scientists but especially have contributed to those primarily interested in biology.

Among the several honors and awards received by Shinya for his professional contributions [Ju-shou Oyobi Eiyo in Appendix 4 of *Through Yet Another Eye*], Japanese readers may be especially interested in the International Prize for Biology that was awarded to Shinya in 2003 by the Japan Society for the Promotion of Science. For this occasion, Shinya and Sylvia were able to bring most of the Inoué family from the United States to Japan and have a reunion with their Japanese relatives. Together they also attended the award ceremony at which the Emperor and Empress of Japan officiated.

In his speech given at the ceremony, the Showa Emperor first congratulated Shinya for his achievements and contribution to science. But as recorded in Appendix 2 of *Through Yet Another Eye,* the Emperor's recollection included how Shinya started his research overcoming difficult wartime problems. That recollection was especially touching for Shinya, who had in fact gone through those trying days. Both the Showa Emperor's speech and Shinya's response at this ceremony (in Appendix 3) can be found under "International Prize for Biology" in Chap. 5.

The balance of his autobiography describes the excitement that Shinya, his students, and colleagues experienced as they further explored the behavior of molecules and their assemblies, which underlie the behavior of living tissues and organisms. Fortunately they continued to unravel hidden secrets of living cells and organisms while exposing the excitement associated with discovery by new biophysical means. Hopefully, their research will resonate among readers and encourage them to explore similar paths of scientific exploration.

As to the interaction of my family with Shinya's, they tended to be private. When in the fall of 1981, we decided to move from Princeton to the Marine Biological Laboratory in Woods Hole, my family remembers how Dr. and Mrs. Woody Hastings invited the Inoués to join the dinner party at their home. This was our first interaction with the Inoués. The Inoués' youngest son, Ted, was about the same age as our own youngest son, Tsutomu, and both being interested in computers were friendly with each other until they entered the university. Our daughter Sachi had a summer job in Shinya's laboratory, and we recall how she told us about learning about feeding the sea urchins and how to carefully prepare the glassware in which to grow the urchin larvae.

In 2002, I retired from the Marine Biological Laboratory in Woods Hole myself. On that occasion, Shinya arranged a memorial symposium in my honor. With little knowledge on my part, Shinya had assembled those scientists interested in aequorin and green fluorescent proteins (GFPs). Those scientists included Roger T'sien from California, with whom I shortly shared the Nobel Prize in Chemistry. The audience filled the available seats and surprised me with the breadth of Shinya's scholastic knowledge and acquaintances. At this meeting, Prof. Ford from Rutgers University, who a couple of years later in 2004 co-organized the Friday Harbor Lab 100th Anniversary Symposium, also joined us. The topic of the symposium was "calcium-dependent bioluminescent protein and green fluorescent protein." Shinya was present at this symposium and reported up-to-date information on the fluorescent polarization of GFP protein crystals, which I had purified for his polarized light microscopy. For this meeting at the Friday Harbor Labs, Shinya's family and we occupied adjoining cottages. And I recall how Mrs. Inoué, who had been raised in the state of Washington, had thoroughly enjoyed the island by driving her car to various points of interest there. That same summer of 2004, a memorial dinner was held at the Marine Biological Laboratory in Woods Hole to commemorate Shinya's contributions to science, and we too were invited for the occasion. Shinya's friends from long distances joined the occasion, which was most appropriate considering his having received the International Prize for Biology the previous year. At this meeting, I remember the many representatives of the world's top microscope companies who recalled the severe demands from Shinya, who challenged them with improvements of the microscopes which they were barely able to satisfy by exercising their utmost efforts.

Since those days, we met on occasions when Shinya was the recipient of various prizes and on the occasion of me receiving my Nobel Prize. With each of these occasions, I have been impressed by how many individuals around the world loved Shinya as a humanistic researcher. As described in his autobiography, not only those in Japan and the United States but research collaborators appear from around the globe. To me, this is an indication not only of Shinya's clear observing power and sharp intellect but that of his human attractiveness. I am confident that the strength of character of Shinya reflected in this autobiography will further benefit the character of international students who digest this essay.

Woods Hole, MA, USA
August 2015

Osamu Shimomura

Preface

I realize there are many stories that can be told by those who came to the United States under differing circumstances: those who sought religious freedom, those tattooed with serial numbers from Auschwitz, those brought to provide labor, and those who sought a better life. They each contribute to a divergent worldview colored by their experience yet contribute in their own ways to the development of the New World and hopefully to the eventual betterment of all mankind.

Arriving as I had, 3 years after the end of World War II, with passport number 50 from postwar Japan, and now having had the privilege for over half a century of participating and observing the evolution of science and technology through the eyes of a biologist interested in how cells work, I may have some stories to tell that could be of interest, at least to some of my family and friends. The yet-untold story of my father's role in the history of the Anti-Comintern Pact that preceded the Pacific War also may be of interest to some since it portrays a facet of Japanese politics prior to World War II.

In Part 1 of the following, *Through Yet Another Eye*, I recall my early childhood in the free world followed by my less wholesome experiences in Japan as a student during the war. By the time Japan attacked Pearl Harbor, I had met Katsuma Dan (Katy), and a new life of hope in scientific research had started to percolate in me. From Part 2 on, I describe my life as a scientist/teacher after arriving in the United States in 1948. Here I had the privilege to extend, and share, my interest in uncovering the behavior of living cells undergoing cell division and to unravel the function of the underlying dynamic molecules. At each step, these biological discoveries required the development of special microscopes, which made the biological research all the more interesting for me.

Some of my stories should not be very difficult to follow. I made an effort to write this whole story of my experience, explaining the observations and discoveries in reasonably plain language. However, here and there I may have slipped into technical jargon when I talk about the optical tools that I developed or when they relate to the behavior of living cells or to their underlying molecules. I trust that you will not be distracted by the technical terms or expressions used in those parts and will manage to move forward and share in the exciting stories of the technical and scientific discoveries.

Acknowledgments

In recollecting the events and episodes covering more than 80 years, I have been helped by recalling my interactions with the many friends, teachers, and others I met in different countries and institutions at various moments in my life. I owe much to them, as well as to members of my families, both in Japan and in the United States, for the inspiration, friendship, and kindness received. Relating to specific episodes, my sisters in Tokyo, especially Futaba Koike and Sumiko Fujii, and my niece Fumi Tsukahara, as well as my daughter Heather Russell; sons Jon, Chris, Steve, and Ted; and granddaughters Katie, Penelope, and Elizabeth in the United States, have provided valuable input.

Throughout the preparation of this manuscript, Chris and Ted, as well as my secretary Jane MacNeil at the Marine Biological Laboratory (MBL) in Woods Hole, have been most helpful in providing advice concerning the organization of the whole presentation, grammatical and spelling corrections, and choice of expressions and illustrations. To them and especially to my wife Sylvia, who has patiently supported my whole effort throughout the writing of this volume and who shared in many of the episodes described therein, I express my heartfelt thanks.

In preparing the English of versions of *Through Yet Another Eye*, two members of our research group at the MBL, Tomomi Tani and Issei Mabuchi, provided essentially all of the translation from the Japanese draft into English. At the same time, Ichiro Sase of Nikon also contributed to the drafting and translation of the Japanese version into English.

And last but not least, we greatly appreciate the editorial support of Springer in Tokyo and the generous financial support of Nikon in Yokohama for the publication of *Pathways of a Cell Biologist*. We are truly grateful for their invaluable assistance.

Woods Hole, MA, USA
December 2015

Shinya Inoué

Contents

List of Figures

Chapter 1
Early Years

Prologue

The Last One to Go

August 15, 1945. The echo of the two bombs reached the far hamlets of the Japanese Archipelago 6 days after Nagasaki. The village doctor, farmers, trades people, students in 'kinro-hohshi' (labor service), and soldiers in uniform all huddled around the radio set to catch the 'important announcement.' Finally the shaky but occasionally firm voice of the Emperor came on the air, a first for the Japanese people for whom he was considered a Living God.

Tears flooded the eyes of most, and some in uniform muttered "mou-shi wake nai" (I failed to live up to my duty), as the Imperial voice declared Japan's surrender to the Allied Forces and implored his people to start anew "bearing the unbearable."

The bamboo spears, with which the populous were "to defend the nation to the last woman and child," were cast to trash heaps, and some of the military officers expressed their remorse for "failing his Majesty" by taking their own lives by seppuku (committing suicide by running a sword across one's abdomen and exposing the entrails – a method used by Japanese samurai to express his remorse for not fulfilling his duty). In a few days, the trains, buses, and trucks, some stacked with booties gained from the Imperial Army stockpiles, were full of those with mixed emotions returning to the cities or to their farms.

In a farming village of Nagai on Miura Peninsula, which makes up the southern arm of Tokyo Bay, the soldiers who were billeted at Katsuma and Jean Dan's home also received the Imperial edict through a tiny radio placed on the Dan's living room window. After some time of stunned silence, "Now it is your turn to teach us how to behave, Dan-sensei" remarked a uniformed officer. Dan was a young biology professor, who had married Jean Clark, a fellow graduate student at the University of Pennsylvania in Philadelphia and at the Marine Biological Laboratory in Woods

S. Inoué, *Pathways of a Cell Biologist*, Springer Biographies,
DOI 10.1007/978-981-10-0947-1_1

Fig. 1.1 Katsuma and Jean Dan with oldest son Teruki (Courtesy Dan family)

Hole, Massachusetts. "Since we followed at the end of the line so far, and now we are to 'about face' so, yes, Jean and I are indeed at the front of the line" was the gist of Dan's response to this young officer.

Ten years earlier, in 1934, Katsuma Dan (Fig. 1.1), whom I shall often refer to as Katy (a frivolous nickname he gave himself) returned from the United States with his young bride Jean and set up a laboratory at the Misaki Marine Biological Station facing Aburatsubo Bay at the far tip of Miura Peninsula. Except for the days when they were in Tokyo, the Dans would bike together from their home in Nagai to the lab, some 2 miles up and down the hills that the buses running on wooden distillates could barely climb. Those happy days lasted until the middle of the Pacific War, when Jean, as an enemy alien, was 'recommended' to move to Tokyo instead of remaining in Nagai. Nagai was part of the fortified zone of Miura Peninsula, and Yokosuka, the main base for the Japanese Navy, lay between Misaki and Tokyo.

Katy's commute to the Misaki Lab and his studies on cell division continued until the Lab was taken over by the Japanese Navy as a miniature submarine base in January, 1945. When the Pacific War finally ended and the Allied Occupation Forces arrived, Dan was itching to return to the Lab to resume his studies. But to the Allied Forces, this was an enemy naval facility. How was Dan to convince them otherwise?

September 28, 1945. Officers from the defeated Japanese navy sat rigidly across the table from representatives of the occupation forces. Occasionally, one or the other of the frightened participants would shout something, totally unintelligible to those across the conference table at the Misaki Station.

No one touched the asparagus and beer provided for the occasion except Katsuma Dan, who was amused by the chaos he observed from the end of the table (Fig. 1.2).

Jan. 16th, 1946

Dear Don:

Hello Don! No matter how many years I havn't written to you, still I have to start in the usual way. How are you? One day toward the end of last year, Lieutenant Coe dropped in, bringing your message. He came all the way from Tokyo in a jeep and must be quite tired, for he didn't stay long. But we gathered enough information about you and we were quite happy.

We got another friend here. Can you guss who it is? Edwin Helwig. But since he has been away from home for three years, the informations we can get through him are slightly stale. ....

In the last four years, we simply had lots of things. But ducking under bombs was not so bad. Rather it was a great excitement. Hide and seek at the expense of your life cann't help being exciting. There was, however, an awful side to it too.

During first three years of the war, our everyday life was not fundamentally affected and we could work regularly. Then, in January, a year ago from now, the Japanese navy took over the station and changed it to a temporary submarine base. As a result, the laboratory moved to a near-by village. Although, we succeeded in getting a wooden building built, we had a hard time to install sea-water pumps and so forth. When the war was over, the American troops occupied the station, for the reason that the Japanese navy had been there. When American officers came there for the first time to take over the place, I went there. It was a funny experience. On one side of a long table, three American officers were sitting and on the opposite side two Japanese navy officers and I were sitting. They served beer and canned asparagus with tomato ketchup. This slightly cute menu made me smile. But, oh boy, the both sides were pretty much excited. I am sure they were really scared of each other. They yelled whatever they wanted to say at the top of their voice but never listened to the other side. And an interpreter translated off and on, paying no attention whether it made sense or not. I was partly absorbed in watching the chaos and partly in the asparagus and was still partly absorbing the beer. Neither side understood the other. But to start with neither side knew what they were going to say. Toward the end, I was loudly laughing which nobody noticed. Somehow the meeting came to end. So I started to work. I stuck to a major and explained to him that this building originally belonged to the University etc., etc. In ten minutes, he began to see the situation. As soon as I saw the sign of dawning in his chaotic mind, I ran back to the building, wrote a psoter asking soldiers to take good care of the place because it is a research institute, pasted it on the wall and took leave from the back door leaving the noisy bunch there. There was no way to know what influence the poster excerted on the American soldiers who came there. But to my greatest surprise, in a copy of over-sea edition of Time issued in the beginning of December, I came across my own words all printed. Moreover, it had a t'tle "Appeal to Gothe".

....

From Katsuma

Fig. 1.2 Katsuma Dan to Don Costello (Courtesy K Dan)

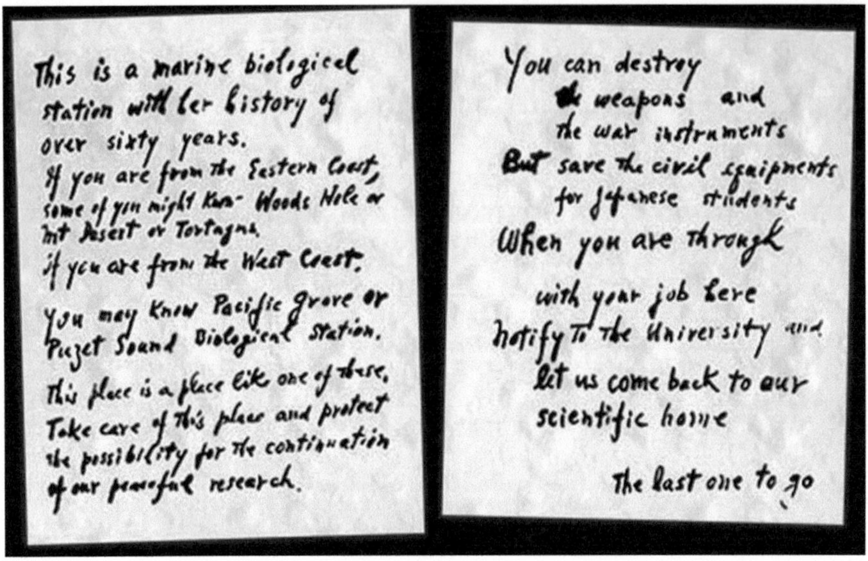

This is a marine biological
station with her history of
over sixty years.
If you are from the Eastern Coast,
some of you might know Woods Hole or
Mt Desert or Tortugas.
If you are from the West Coast,
You may know Pacific Grove or
Puget Sound Biological Station.
This place is a place like one of these,
Take care of this place and protect
the possibility for the continuation
of our peaceful research.

You can destroy
the weapons and
the war instruments
But save the civil equipments
for Japanese students
When you are through
with your job here
notify to the University and
let us come back to our
scientific home

The last one to go

Fig. 1.3a "The last one to go" (Inoué S, A Tribute to Katsuma Dan, Biol Bull 187 (2): 125–131, 1994. [A 56])

After this inconclusive meeting, Dan posted a message that he had hurriedly composed on a piece of brown wrapping paper in brush and Chinese ink, on the Lab door. He signed the message "The last one to go" (Fig. 1.3a).

Both of these two messages were sent by Captain Parks (Fig. 1.3b), via the Woods Hole Oceanographic Institution (WHOI), to the Marine Biological Laboratory (MBL). All three of these messages are now prominently displayed in the hallway to the MBL Library.

On the last day of 1945, Dan was finally summoned by Lieutenant Major (**?**) Spooner of the U. S. First Cavalry and handed the document releasing the Misaki Marine Biology Station (Fig. 1.4) back to the University of Tokyo (Tohdai).

The following summer, Dan's laboratory at Aburatsubo started humming again, joined by his former students Yoshiyuki Endo and Tomonao Ohno. Those were the two who had paid an excited visit to Katy's Nagai home exclaiming "free at last," the day following the Imperial edict. For me, who had been conscripted into the Imperial Cavalry for the last 4 months of the war and was suffering from malnutrition, it was yet another year before I could rejoin this jubilant group.

I started this narration by portraying the traumatic final day of WW-II and then by introducing Katsuma (Katy) and Jean Dan. But I have left out how the War impacted on my own school days in Japan, from grade school through university, and how that all changed with my meeting with the Dan's. I shall now describe those topics in the following sections.

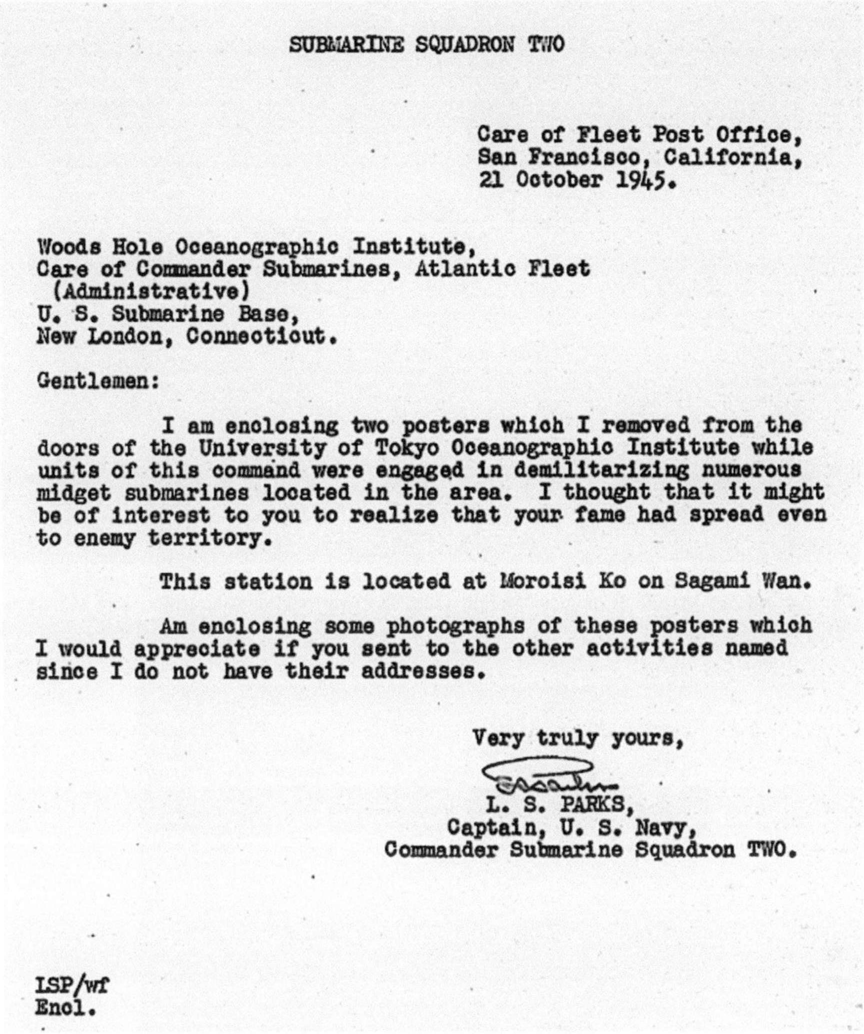

SUBMARINE SQUADRON TWO

Care of Fleet Post Office,
San Francisco, California,
21 October 1945.

Woods Hole Oceanographic Institute,
Care of Commander Submarines, Atlantic Fleet
(Administrative)
U. S. Submarine Base,
New London, Connecticut.

Gentlemen:

I am enclosing two posters which I removed from the doors of the University of Tokyo Oceanographic Institute while units of this command were engaged in demilitarizing numerous midget submarines located in the area. I thought that it might be of interest to you to realize that your fame had spread even to enemy territory.

This station is located at Moroisi Ko on Sagami Wan.

Am enclosing some photographs of these posters which I would appreciate if you sent to the other activities named since I do not have their addresses.

Very truly yours,

L. S. PARKS,
Captain, U. S. Navy,
Commander Submarine Squadron TWO.

LSP/wf
Encl.

Fig. 1.3b Letter of transmission by Captain Parks (Courtesy MBL)

Shadows of War

Childhood

After returning to Tokyo from London, England, where I was born in 1921 (Fig. 1.5), and Quemoi, China, where my father had worked in the Japanese Consulate, I was now old enough to start grade school. But the first day was a complete fiasco. Not

Fig. 1.4 Misaki Marine Biological Station (MMBS) overlooking Aburatsubo Bay (Courtesy MBL)

Fig. 1.5 Shinya in London (Courtesy of Inoue family)

Fig. 1.6 Family at Long Beach. From *top to bottom*: mother, father with Akiko to his *left* and Midori to his *right*, Shinya, and Sumiko (Courtesy of Inoue family)

wanting to go and meet a totally unfamiliar group of youngsters in Japan, I had hidden, bawling in the basement of the home our grandfather had built for us in Magome, until I was dragged out and summarily sent to the local grade school.

Fortunately within a month, my father received his next posting, this time to Portland, Oregon (Fig. 1.6). Unlike the fear I had felt for unknown experiences in a Japanese school, I enjoyed life in the States, including Ainsworth School located not far from our home on Rose Heights in Portland.

I can still recall the story of the kingfishers told by our teacher. Also, I remember the garter snakes that Anne, a 5-year old from the family across the street, would wind around her wrist and the salamanders that she and I would catch in the brook behind her house. For some reason 'Idaho potatoes 1928' still rings in my ears after hearing the jingle (on the radio) on a 3-day-long trip to Yellowstone Park with my father and his friend.

Mother used to take us to the shows in the department store where miniature pictures of performers appeared in a foot-diameter window. To this day I can't figure out if those were early TVs or some optical effect produced by a large negative lens. In those innocent days, I was not aware that the friendly barber, who would come and cut our hair at our home, had told my father, "It's just that the other customers do mind if you come to my shop."

After 2 years in Portland, we moved to Sydney Australia, where my father was Consular General. I attended Cranbrook School at the base of Point Piper, where our residence was located, while three of my four sisters were driven to Kambala School at the other end of Rose Bay. My proud accomplishment at Cranbrook was the class's best English composition that I wrote soon after arriving in Sydney. That naturally helped boost my confidence. The cold showers after rugby practice on early winter mornings and the mutton and rabbit stews offered weekly at the school dining room also linger in my memory.

Our house, 'Craggy Moore,' overlooked Sydney Harbor, where the Harbor Bridge was being extended from the North and South shores daily, until 1 day they met above the bay to complete the span. As commemoration for the occasion, delegates of the foreign services were given an airplane ride above the harbor, a rare privilege in 1932. It was also great fun for us youngsters, but unfortunately not so for my mother for whom a plane ride was too much of a new experience. For souvenir, I was given an inch-thick section of one of the 4-inch-diameter cable bundles that had braced the growing ends of the bridge until they met in the middle.

It was in Sydney that one of my classmates, a Dunlop who lived near our house, introduced me to crystal radio. Winding our own coils, adjusting the variable capacitor, and probing for the best spot on the 'crystal' with 'cat's whisker,' after connecting the coil to the ground rod on one end and to a long antenna insulated on the other end, the broadcast came through the head phones on lucky days. My parents also gave me my first Kodak camera, whose film cartons I assembled to build my own model of Sydney Harbor Bridge.

Another blond-haired friend, Phillips, was the son of a British Admiral. In the early War years, some 10 years later, I heard the sad news that Admiral Phillips declined rescue and went down with his battleship, the 'Prince of Wales' which was sunk by the then imperious Japanese Navy. But while we were in Sydney, the War was still far away in the horizon, and the young Phillips and I had fun innocently attending each other's birthday parties at the Admiralty and the Consulate.

Those were the last of my happy school days for several years to come. From Sydney we returned to Japan, at an age when I should have entered sixth grade. But already competition for entrance to high school was fierce in Tokyo. So I had to repeat fifth grade at Magome Dai-ni, not far from our home, to catch up on my Japanese in a year and a half. Fortunately my grade school teacher, Mr. Moriyama, a man of small build but full of cheer, was an enthusiastic and sympathetic teacher. So after his tutelage, I managed to be accepted to the newly opened Second Tokyo Municipal High School in Ueno Park. The commute was a 30-min train ride between Oomori and Ueno stations, sandwiched by 15-min walks at each end of the ride.

High School Years

Of the five adolescent years of schooling in Tokyo, I have not too many good experiences to remember. Japan had invaded Manchuria and then China and the high schools were becoming more and more militaristic. Strenuous exercises and rote learning of nationalistic history and Chinese classics were emphasized, and the English classes were taught by teachers who had little first-hand experience with the language.

I found the 'Japanese English' stilted and difficult. My red necktie was completely out of place, and my hair had to be sheared to a quarter inch of my scalp (Fig. 1.7).

Our headmaster idolized Admiral Yonai who was from his home town and encouraged the best students from our classes to apply to Etajima, the National Naval Academy. A decade later, many Etajima students including our class valedictorian, Akaba, were the first to perish in the Battle of the North Sea.

Still, I remember a few of my high school teachers warmly. Mr. Seki, a botanist, introduced us to microscopes, first using simple magnifiers on adjustable stands. Through the magnifier, I still remember marveling at the tiny zippers that nature had built on to bird's feathers, which allowed them to be combed apart or to be fastened back together in neat sheets. After classes, a graduate from our high school introduced us to shop, helping us build steam engines and other functional models. Using 'peanut' vacuum tubes imported from RCA, I also managed to build a three-tube

Fig. 1.7 Adolescent Shinya at Magome (Courtesy of Inoue family)

radio set, for which I won first prize at the City Electrical Museum. The spider coil served as an antenna, which was a fine innovation, but neither I nor the examiners caught on to the radio-wave insulating properties of the pretty, tin soap box into which I had assembled the set.

And then there was the National Museum of Science in the Ueno Park, next to the Music Academy, which all four of my sisters eventually would attend, located between our high school and the train station. That is where I first came across television, in which streaks of light were cast from a spinning Nipkow disk onto the performer whose image appeared on a remote Brown (oscilloscope) tube. The technology was fascinating, even though I could not understand why a dozen or so photoelectric cells were used to capture the light scanning the performer. These were the days before the imaging tube (iconoscope) was invented, so the subject was scanned by a beam of light rather than an electron beam scanning the image formed on the photo cathode of the imaging tube (see e.g., Inoué, 1986, Chap. 1).

On another visit to the National Museum of Science, I was trying out a self-testing exhibit of the Weber (or Weber-Fechner) Law. What the museum had provided was a series of weights of different sizes. From these weights, we were to determine which were of similar weight by holding one in the left hand and another in the right hand. According to Weber-Fechner's Law, one should be able to tell the difference between two weights when one is a certain percentage heavier than the other [just as in vision, where one can tell which part of an object, or image, is brighter when there is a 2 % difference in brightness (assuming there is no border line between the two areas, that there is enough light, and that the areas of the image are not too small)]. It turned out, just as Weber-Fechner's Law said, that when one weight was heavier than the other by, say, one fifth, regardless of their absolute weight, one could feel the difference and tell which was heavier.

So far so good. But that day I had in my canvas brief case a bottle of concentrated nitric acid that I had acquired from our school's science lab. So when one of the heavier weights rolled and hit my brief case, crunch went the bottle and my brief case and pants got soaked with the acid. I dashed to the rest room and washed off the chemical as best as I could, then rushed back home. But the acid was so strong that it ate right through my brief case as well as my clothing. My parents were away in Germany at that time, and I had to tell my aunt Mitsu what had happened, but I don't recall how she reacted. Perhaps though, that episode explains why my grandfather Kiyomatsu got so upset when, some days later, I brought home several dozen test tubes filled with an assortment of reagents, again acquired from the science lab at school.

A few years before that episode, I had tried many times to build small electric motors that would spin successfully. The relation of magnets to electrical current was most captivating, and the use of a commutator to switch the magnetic field depending on the orientation of the rotor seemed so ingenious. But, alas, I was not aware of the fragile nature of the insulation surrounding the enameled wire. The result was that there were occasional 'shorts' between the insulated wire (that I had unwound and reused from old transformers) and the yolk (which I had made from layered strips of tinned iron sheets that I had cut from an old tea can). This meant that my motors were repeatedly a hit or miss; some would turn for a while when connected to fresh batteries, but others would just hum and not turn even when the shaft was given a twisting nudge! Despite not having warned me of possible shorts caused by loss of the enamel insulation, the magazine, '*Kagaku toh Mokei*' (Science and Models), was a constant companion for me during my efforts to learn how electrical and mechanical devices worked.

While cosmopolitan, to the point of encouraging my sisters to master music as an international language, my parents were still loyal Japanese subjects. They probably did not realize that my negative experience in the Japanese high school would eventually drive me to leave Japan and go to the United States for good. But that is still a long way off.

Musashi

After high school, I was fortunately admitted to Musashi Kotoh-Gakko, a private 3-year liberal arts college preparatory to University. That was in 1938, shortly before my parents and three of my younger sisters went to Budapest, where my father served as Japanese Minister to Hungary, Yugoslavia, and Czechoslovakia. His stationing was prompted in part by his disagreeing with the foreign minister and the military regarding the Tri-lateral Axis Pact between Berlin, Rome, and Tokyo. More on this episode later.

At Musashi, the Headmaster was a philosopher, who advocated our future contribution to spanning the culture of the East and West. I still recall the following story that he told us, a touching little tale but perhaps especially significant in that it was related to us during the rising tides of anti-Western militarism. "A young American girl, perhaps only 5 or 6 years old, was playing with her mother's gold coin. Somehow it had slipped from her hand and rolled away, never to be found again. Years later, when the girl was ready to earn her own way, one of her first purchases was a gold coin, which she presented to her mother with an apology for losing hers as a young girl."

Again at Musashi, I was not an A student. I flunked my tests in higher algebra and in German, and, I was told, the professors debated whether I should be advanced to the next grade or not. Many years later, I learned that two of my best friends, Tsuruta and Tamiya who ranked at the top of the class, had approached the algebra teacher and pleaded for my promotion by sharing their superior grades with me. But they were scared off by being told that if they persisted, their own promotion would also be placed in jeopardy. As it turned out, apparently one of my German teachers had convinced the Committee that it was OK for me to be promoted.

Speaking of German teachers, there was Frau Kanokogi, a Polish lady who had married a well-known Japanese philosopher and then since divorced. Her teaching which I appreciated was not mainly through grammar, unlike the many other foreign language teachers that I had encountered in Japan (whose own learning was primarily through books). Instead hers was through conversation, songs, and actual usage. She was fluent in many languages, even taking the roll call using the list of Japanese names written in Chinese characters, a rare accomplishment for anyone not brought up in Japan. I liked her very much, but she was apparently not involved in my promotion discussions at Musashi.

Entropy

Another class at Musashi that I remember is one in Physics. There, the professor was introducing us to entropy by jotting on the blackboard how to calculate entropy changes from differences of free energy and absolute temperature. The equation was simple enough, but its meaning was a complete mystery to me. I asked for an explanation but was told that the equation itself defined entropy. Frustrated, I pursued further by challenging the teacher, "I bet you're going to ask us to calculate entropy in our exam" to which he replied in the affirmative. I went further by declaring that I wasn't going to answer such a question and followed my declaration by deed.

Such cheeky behavior! But it did take several years before the meaning of entropy finally sunk in as I was reading a paper by Ryo-Go Kubo on elasticity of rubber. If I recall correctly, he pointed out that each carbon-carbon linkage in the molecular chains in rubber can take on only one of three alternate tetrahedral angles. When rubber is stretched, the molecular backbone is straightened out, thus reducing the alternate number of curved zigzag shapes that the backbone can take on. In other words, the entropy for the chain of rubber molecules is reduced by work done on it by stretching. When the rubber band is released, the sequential carbon-carbon linkages become thermally randomized, thus, the backbone shortens, releasing mechanical energy and absorbing heat. In other words, rubber contracts by increasing its entropy. I was delighted to be able to feel on my lips the heat evolved when I stretched a strip of rubber balloon and then the cooling (absorption of heat) when the stretched balloon was suddenly released.

At Musashi, our Chemistry teacher, Dr. Tamamushi, was a well-known physical chemist for whom the Ne-Zu Foundation, which supported the College, had built a special building for his research. One day my friends, Tsuruta and Kamiya, and I approached Dr. Tamamushi to see if he would let us try some project in his lab. He suggested that we measure the rate at which suspended clay sediments in a cylinder of water. A seemingly dull experiment that we soon gave up, not realizing that he was, in fact, introducing us to the sedimentation velocity of colloidal particles from which we should have been able to derive their particle weights!

Introduction to Misaki

I was first introduced to the marine lab at Misaki in 1941, while still a student at Musashi. Carrying our ration of rice needed for our week's stay together with the school's microscope in our backpack, we arrived after an hour-and-a-half train ride to Yokosuka, then by bus to Hikihashi, and finally a half-hour walk amidst rice fields to the lab. At the Lab, seated by the windows of the semicircular room overlooking Aburatsubo Bay, the northern skylight helped illuminate our microscopes.

It was in that Bay where we were introduced to the rich plankton, after learning to skull with the long, single oar on a fishing boat. We learned to row slowly so that we could properly tow the plankton net along the waters where their colors changed at the border of the warmer Kuroshio and cooler current from the North. Much to our delight, we even caught a yard-long chain of *Salpa*, a colonial tunicate whose individuals were nearly 6-inches long.

On the boat-launching ramp, slippery with green algae, we saw long trains of a dozen or more 'sea cows,' the hermaphroditic nudibranch snails that acted as female to the individual in front and male to the one behind.

Back in the lab, we were introduced to the tunicate's heart, which would repeatedly pump blood through the open vessels in one direction for nearly a minute, and then reverse the direction of beat so that the blood was pumped in the other direction for the next minute or so.

Then there was the delicate but unmistakable envelope that arose from the sea urchin eggs in a few minutes after they were fertilized. "Even the moon is surrounded by a fertilization envelope" is a haiku that some zoology student had composed on his way back to the dorm, after hours of observing fertilization of sea urchin eggs through the microscope.

To such peaceful environs, the sounds of war seemed remote until, on the train ride back to Tokyo, we encountered groups of young soldiers surrounded by their town folks waving them farewell with a sea of Japanese flags, accompanied by the ever present music of Gunkan March.

Introduction to Katy

Encounter

A few weeks after the fall term started in 1941, our zoology teacher Yanagida, who also taught classes at Ichikoh, announced that a new person would be taking over our class at Musashi starting next week. The following week, in walked a short but athletic looking new instructor. He surprised us all, first by perching on the corner of the lectern and then by asking, "What shall we talk about today?" – behavior and utterances beyond imagination by Japanese students of those days. That turned out to be Katsuma Dan, better known as Dan-sensei, or Katy (Fig. 1.1).

On receiving no voiced feedback, he volunteered, "Then maybe I'll talk about some of the research on cell division that we've been doing." The studies he told about were fascinating. Instead of stale recitals of textbook descriptions of prophase, anaphase and cytokinesis, he talked about the change in distances between the kaolin particles that he and his colleagues had sprinkled onto surfaces of dividing sea urchin eggs. By using a camera lucida to trace the position of the kaolin particles adhering to the surface of the dividing cell (freed of its fertilization envelope), they found that the particles far away from the cleavage furrow kept moving away from each other, while those placed around the cleavage furrow, first came closer together then moved rapidly apart from each other as the cell cleaved further. And these unexpected particle behaviors – the differential contraction and expansion of the cell surface during cell division – could be precisely accounted for by a model of the cell that contained two rigid asters which were pushed apart by an elongating spindle.

Here at last we were exposed to science in the making! I must have gotten thoroughly hooked by experimental biology and by Katy himself when, a few weeks later, he suggested some laboratory projects that we might like to tackle. The one that Tsuruta, Kamiya, and I chose was the following: to see what we could make of Lillie's iron wire model. As told by Katy, Ralph S. Lillie at the MBL (the younger brother of Frank R. Lillie, the embryologist who was Director of the MBL and for whom the Lillie Building was named) developed a model of nerve conduction that had uncanny resemblance to conduction of electrical impulses down a real nerve fiber. Lillie's model was made as follows:

Take a piece of thick wire made of ***pure*** iron. Soak it in concentrated nitric acid and watch the silvery wire give off a brown cloud. Even though its color has returned, the iron wire is now passivated, that is, its surface is now covered with a very thin, invisible new chemical layer that exhibits a high electrical resistance. Now 'stimulate' the passivated iron wire with electricity, or by pinching a portion of the wire. A brown ring of 'impulse' then travels from the stimulated region along the wire at a constant speed, in both directions away from the stimulus. What a fascinating model. We couldn't wait to try it out ourselves.

So while Tsuruta and Kamiya ran to the chemistry lab to beg for a bottle of concentrated nitric acid from Dr. Tamamushi, I went out to the road in front of our school with a pair of wire cutters. There I saw a long piece of 2-mm-diameter wire wound around and around to reinforce the telephone pole. Thinking that a short loop of the wire would probably not affect the reinforcement, and that even though this could not possibly be made of pure iron it might work, I snipped off a foot-long section and headed back to the lab.

When our two prizes were put together in a long glass tray, sure enough the wire gave off a puff of brown cloud which quickly disappeared. And when we touched the passivated wire with a pair of forceps whose arms formed a copper-zinc pair, much as the one Galvani used to stimulate the frog nerve-muscle preparation, out came a puff of brown cloud that traveled away from the stimulus, each as a brown ring about a centimeter long. We observed the same change when we pinched the passivated wire with a sharp pair of ordinary forceps, or scraped it with a scalpel. So even with the steel wire from the telephone pole, we were able to repeat Lillie's iron wire model.

The question that arose among us right away was: is the impulse being conducted by a series of chemical reaction, as Tsuruta and Kamiya (both of whom later majored in chemistry at the University) proposed, or as a chain of electrical events (a wave of depolarization) more akin to real nerve fibers as I (who headed for zoology) had argued. By then, Katy was sitting to the side smoking his aromatic pipe and enjoying the youngsters' arguments.

My suggestion for testing our alternate hypotheses, a test to which my contestants agreed, was to make insulating dams out of paraffin, so that we now had two lakes of the acid, with the part of the passivated iron wire spanning the two lakes free of nitric acid. This way the part of the iron wire between the two lakes was not exposed to nitric acid and hence was insulated relative to the acid pools. The brown impulse would then travel as far as the dam but not beyond, unless I argued, we made an external bridge which connected the two lakes of nitric acid electrically.

Much to my joy, the impulse stopped at the first dam, but jumped across the barrier and immediately restarted at the other side of the second dam when the two lakes of acid were connected with a piece of copper wire, or with glass tubing containing electrically conductive acid or salt solution.

Our experiment showed that the naked iron surface, newly exposed by scraping the passivated layer or by breaking down the passivated layer by electrical stimulus, must now have a different surface electrical potential compared to the still passivated part of the wire. In essence, it became the pole of a battery from which current would flow through adjoining, still passivated regions and back through the iron wire itself. That current would then break down the passivated layer adjoining the stimulated region and expose a new naked iron surface while the initially exposed iron surface would eventually become passivated again. In the meanwhile, the newly exposed iron surface would become the source of another current loop, and the process would repeat itself generating a new current and exposing another new region. In other words, Lillie's model propagates a wave of electrical depolarization just as in a real nerve fiber.

In our experiment, we had added an extra insulating region (the dry region between the two paraffin-rimmed dams) across which the impulse would continue, only in the presence of an external electrical shunt. But in the presence of an electrical shunt, the electrical impulse jumped across the barrier instantly, much faster than if it had traveled as a wave of depolarization through the nitric acid solution. So, in effect, we had converted Lillie's iron wire model into a model of myelinated nerve. The axons of myelinated nerve fibers are wrapped by a few millimeter-long sheaths of myelin (an electrically insulating layer made by many, many layers of cell membranes), whose junction, the Ranvier node, does allow electrical impulse to pass to the interior conductive layer of the axon.

Thus, as Dr. Ichiji Tasaki had shown in 1939, the conduction of impulse along the nerve axon is sped up many times by the presence of the myelin sheath compared to naked (non-myelinated) nerve fibers by having the membrane-depolarizing electrical currents jump from one node of Ranvier to the next. To a college sophomore, such experiences of discoveries were fatal. How could one not get hooked into specializing in biological (or biophysical) research! But this was just the beginning of my encounter with Katy and with his wife Jean.

Shortly after Katy started teaching our class at Musashi, he and I walked out to the tiny radio store in front of the school to catch the news. There, we heard the excited voice of the announcer accompanied by the blaring 'Gunkan March' that "Early this morning the Imperial Navy has attacked and inflicted considerable damage to the American Fleet in Pearl Harbor." That was December 8, 1941.

Even though Katy had not been totally oblivious to the potential of such an event [while still a graduate student at the University of Pennsylvania, Katy himself had received the news that his father Baron Takuma Dan was assassinated (as being one of Japan's leading capitalists) by a militarist hot head in Tokyo on March 5, 1932], Katy's depressed utterances were "Taihen Na Koto Ni Natchatta" (this is really terrible). Here he was in Japan with his young American wife, and his own country had started waging war with his wife's homeland, a country with many of their dearest friends.

The Anti-Comintern and Tri-Axis Pacts

Father

Since we are talking about the opening days of the Pacific War, I shall digress here, from relating my own development as a student in science, to touch on my father's experience in the Japanese Foreign Office.

My father, Kojiro Inoué, was the son of Kiyo-Matsu and Ta-Ma Inoué. Grandfather Kiyo-Matsu, who was born in the fifth year of An-Ssei (1858, 'before dawn arrived' to Japan in the Meiji era), had escaped a small village in Kanagawa Prefecture to seek a new life in the international port city of Yokohama. There Kiyo-Matsu carved out a meager living, at times operating a tiny corner tobacco store, or inventing black shoe polish which sadly would never dry, and eventually rising to become a building contractor. My father recalls how ashamed he was even to invite his high school friend to their home, because one could hardly fit standing in front of the urinal in the tiny bathroom of their home.

As contractor, grandfather Kiyomatsu was said to be a task master, who would not tolerate an iota of sloppy work. I am told that, when building the house for us in Magome (after the big Kanto Earthquake had demolished his and all the surrounding homes in Yokohama), three groups of carpenters quit because of his demanding requirements for building what he considered to be a perfectly earthquake-proof home. But then, I am told, one of his proud days was at the anniversary of the opening of the Yokohama High School whose building he had contracted. Because he had made sure that there was no cheap mix in the concrete, the School had stood soundly through the devastating earthquake in Yokohama.

Fig. 1.8 Kojiro and Hideko Inoué (Courtesy of Inoue family)

Kiyo-Matsu and his wife Tama had 9 children, 4 boys and 5 girls. Despite their meager means, they managed to raise their children admirably: the eldest son, Kanichi, became a successful business man in Shanghai; the second, my father Kojiro, rose to ambassadorship in the foreign service, the third, Kinzou, was a Mitsubishi Bank director; and the fourth, Seishirou, became a high school English language teacher. The girls all married into respectable homes.

After successfully completing the exceptionally difficult and competitive entrance exam for foreign-service (especially for those with no societal connections) my father was briefly stationed in Peking, China. Soon he returned home and married my mother, Hideko. He is said to have been most impressed with his wife to be, a second year student at the Tsuda Women's English College, who said that she ***wanted to live in truth***.

To us children, and many others, she is remembered as a most sweet yet intelligent person whose cooking and dessert were world famous. Shortly after their marriage, they traveled to London, England (Fig. 1.8).

While father worked as a young officer at the Embassy, my mother at our home in Ealing had three babies who were born just over a year apart: myself, the oldest and only son, and then Sumiko and Midori, my sisters.

In London, they befriended Dr. Kobayashi and his wife, who was a friend of my mother's from Tsuda Women's English College in Tokyo. Her husband, a jolly physics exchange student, who apparently introduced me to a few of my first Japanese words such as "kisha" (train) and "chokkon" (chocolates), was later to influence my life in many ways.

My parents recall their horror in 1924, when the telegraph machine in the Embassy clattered out "the whole Japanese Archipelago has disappeared beneath the sea due to a major earthquake." Fortunately this was a reporting error, but nevertheless reflected the catastrophic fire and damage done to Yokohama and Tokyo by the big Kanto Earthquake.

Fig. 1.9a L-R: Midori, Shinya, Futaba, Sumiko, Akiko in Sydney, Australia (Courtesy of Inoue family)

Another memory I have of my father is the story he told me about the month-long boat trip we took from London back to Japan, through the Suez Canal and Indian Ocean. He was holding my two sisters in his arms and was unable to catch me when I tripped from the steep stairs that led up to the observation deck. Fortunately, my young body suffered no broken bones, but perhaps the subconscious memory of this episode is why even today rocking boat and train rides, and even the smell of deck paint, make my stomach feel unsettled.

After a brief stay in Japan, our family spent 2 years in Quemoi, China, where my third sister, Akiko, was born; then 2 years at the Consulate in Portland, Oregon; and two more years in Sydney, Australia, where my youngest sister Futaba was born (Fig. 1.9a).

Then, since I was already in high school, I remained in Tokyo with grandfather Kiyomatsu and aunt Mitsu (an unmarried English teacher in a girl's high school, who stayed unmarried to care for her aging father), while my parents and sisters traveled to Berlin.

Father's 'KenPaku-Sho'

During his posting in Germany between 1935 and 1937, it became father's task, as Acting Ambassador, to negotiate the Anti-Comintern Pact with those in the German Foreign Office under Foreign Minister Joachim von Ribbentrop. Attending the

Fig. 1.9b Sumiko (*left*) and Akiko (*far right*) clapping to encourage Japanese swimming champions at the Berlin Olympics (Courtesy of Inoue family)

Berlin Olympic Games during the daytime together with my mother and sisters (who cheered for the Japanese athletes new to the Games Fig. 1.9b), father and Officer Furuuchi worked secretly in the Embassy composing coded telegrams late at night until daybreak. Having heard that the Japanese Army's telegraph codes may have been broken by British Intelligence, father sent the Embassy messages to the Tokyo Home Office using the Foreign Office's own code.

The Anti-Comintern Pact, initially agreed upon between Germany and Japan, was a mutual commitment to prevent the spread of Communism and, in a secret Addendum, to provide mutual military aide against the Soviets if the Soviets invaded other countries. In 1937, Italy, and later Hungary, Manchuria, and other countries, joined the Pact.

After Ambassador Mushakoji returned to Berlin, the Anti-Comintern Pact and Supplementary Protocol were officially signed in Ribbentrop's office on November 25, 1936. Photographs for that occasion (Fig. 1.10), with my father (second from right) looking over the Japanese and German Ambassadors signing the documents, have appeared on history television and in U. S. text books, sometimes mistakenly attributed to the signing of the Three-Axis Pact.

Father would have turned over in his grave had he seen those wrong attributions, since he had fought hard, although sadly in vain, to prevent the Anti-Comintern Pact evolving into the Three-Axis Military Pact.

Fig. 1.10 Signing of the Anti-Comintern Pact (Courtesy Japanese foreign service)

But even during preparation of the Anti-Comintern Pact, the Military Attaché to the Japanese Embassy in Berlin, General Ooshima, was conspiring with German military personnel to expand the Pact to include a military alliance between Berlin and Tokyo, not only against the Soviet Union, but also against England and France.

While Acting Ambassador to Germany, father had prevented the inclusion of such far-reaching agreements, but when he returned to Tokyo and was put in charge of the Asian and European Division of the Japanese Foreign Office, he had greater and insurmountable battles to face.

On March 23, 1939, the Foreign Minister, Ari-Ta, who had just returned from the five ministers' meeting (Prime Minister, Army, Navy, Foreign, and Treasury), told father that he had reluctantly signed the agreement that Japan accept the responsibility 'in principle' to cooperate with Germany and Italy in case of war against England, France, and others, not only the Soviets.

Ari-Ta told father that now that Naval Minister Yonai had given up his opposition, and that Ari-Ta was therefore left in a position of four to one, he could no longer hold out against the other ministers. Father speculates that Yonai changed his position after being harassed by his junior naval officers; overpowering of seniors by the juniors being rather common in the Japanese Military during those days.

Seeing that the Foreign Minister had fallen in with the military, father prepared a handwritten 'KenPaku-Sho' (letter of plea; Fig. 1.11) that evening addressed to Prime Minister Kiichiro Hiranuma. In the 'KenPaku-Sho,' father argued that

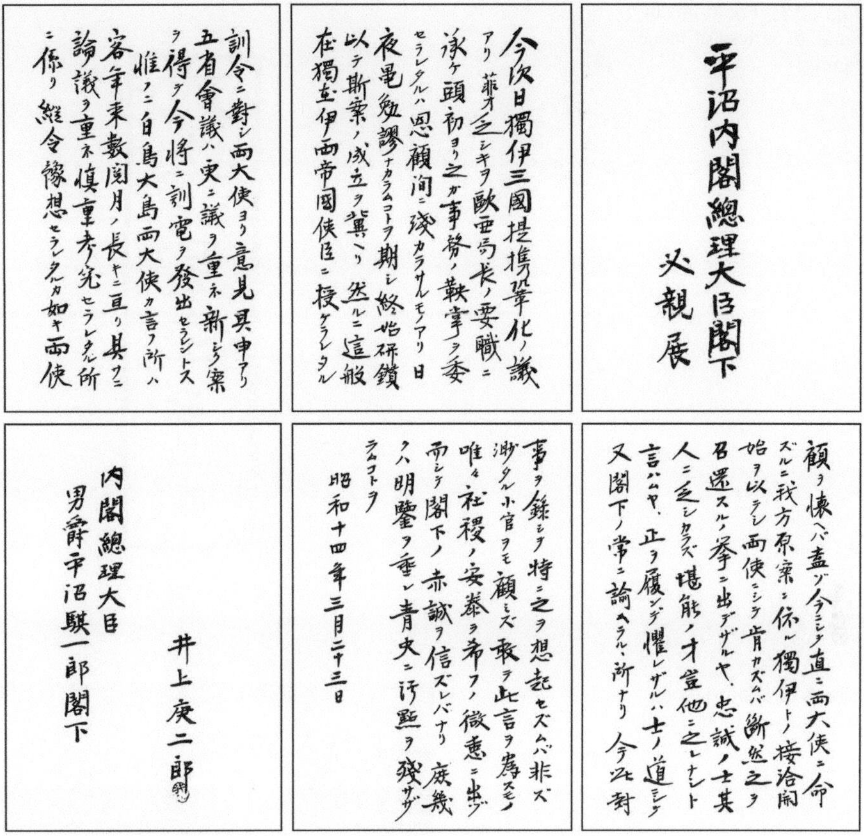
平沼内閣總理大臣閣下
必親展

今次日獨伊三國提携強化ノ議アリ菲才之ニシキヲ歐亞局長ノ要職ニ泳ケ頭初ヨリ之カ事務ノ執掌ヲ委セラレタルハ恩顧洵ニ淺カラサルモノアリ日夜黽勉謬ナカラムコトヲ期シ終始砕鑽以テ斯案ノ成立ヲ冀ヘリ然ルニ這般在獨並伊兩帝國使臣ニ授ケラレタル

訓令ニ對シ兩大使ヨリ意見具申アリ五相會議ハ更ニ議ヲ重ネ新ニ案ヲ得テ今將ニ訓電ヲ發出セラレントス唯フニ白鳥大島兩大使カ言フ所ハ客年來數閲月ノ長キニ亘リ具サニ論議ヲ重ネ慎重考究セラレタル所ニ係リ縦令豫想セラレタルカ如キ兩使

……

顧ヲ懐ヘバ盡ク今ニシテ直ニ兩大使ニ命ズルニ我方原案ニ依ル獨伊トノ接洽開始ヲ以テシ兩使ニシテ肯カズムバ断然之ヲ召還スルノ挙ニ出デザルヤ忠誠ノ士其人ニ乏シカラズ堪能ノ才豈他ニ之レナシト言ハムヤ正ヲ履ンデ懼レザルハ士ノ道ニシテ又閣下ノ常ニ諭ヘラルヽ所ナリ今此封

事ヲ録シテ特ニ之ヲ想起セズムバ非ズ渉タル小官ヲモ顧ミズ敢テ此言ヲ為スモノ唯々社稷ノ安泰ヲ希フノ微衷ニ出ヅ而シテ閣下ノ赤誠ヲ信ズレバナリ庶幾クハ明鑒ヲ垂レ青史ニ汚點ヲ殘サザラムコトヲ

昭和十四年三月二十三日

井上庚二郎

内閣總理大臣
男爵平沼騏一郎閣下

Fig. 1.11 Father's *'KenPaku-Sho'* to Prime Minister Hiranuma. [This figure shows the first and last three pages of the *'KenPaku-Sho,'* handwritten by father in classical Japanese by brush and ink. The full text of this heartfelt document spans 20 pages. (As in all classical Japanese and Chinese text, it is read from the top down, right to left.) The full text of the *'KenPaku-Sho,'* together with its official introduction and their English translations, are given in Appendix I] (Courtesy of Inoue family)

Ambassadors Ohshima in Berlin and Shiratori in Rome both overstepped and disregarded their charges from the Foreign Office (of which father was in charge of the Asian and European Division) and that the new situation placed Japan in an untenably precarious position, even in the worst case, obliging her to fight against the United States.

The next day, father hand delivered the 'KenPaku-Sho' to Prime Minister Hiranuma at the Diet Building. But after a hurried glance through the message, Hiranuma returned the document without comment. Furious, father mentions in his autobiography '*Sajo no Ashiato*' ('Footprints in Sand,' Fig. 1.12; Father, Fig. 1.13) that he could have stabbed the Prime Minister on the spot, had he a weapon with him. With nowhere else to turn, father reports that he left his future disposition to his immediate boss, the Foreign Minister Ari-Ta.

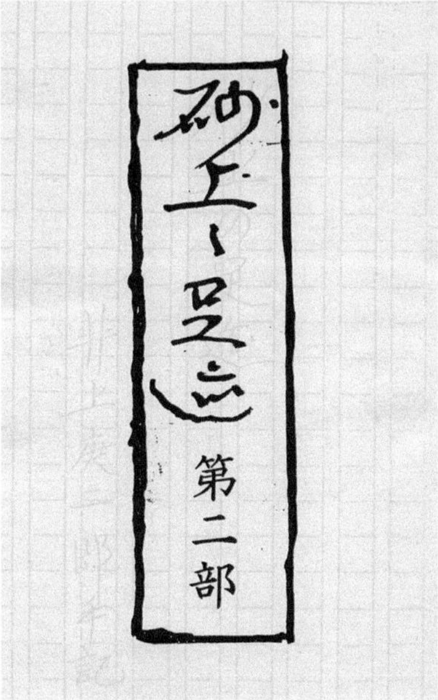

Fig. 1.12 'Footprints in Sand' (Courtesy of Inoue family)

Fig. 1.13 Kojiro Inoué (Courtesy of Inoue family)

Following the ascendancy of Yohsuke Matsuoka to Foreign Minister, the Three-Power Military Pact between Germany, Italy, and Japan was signed in Berlin on September 27, 1940.

In August of that year, the Hiranuma Cabinet was dissolved upon learning the 'inscrutable announcement' that Germany had signed a mutual non-aggression pact with the Soviets. This new pact directly contradicted Article Two of the Secret Addendum to the Anti-Comintern Pact in which Germany and Japan were not to undertake any action inconsistent with the Anti-Comintern Pact. Dissolution of the Hiranuma Cabinet gave the Japanese Military and its sympathizers the opportunity to gain even stronger hold on the Nation's policies.

By that time, father was posted to Hungary, Czechoslovakia, and Yugoslavia as Minister, apparently in part to remove him from positions of influence in Tokyo. He was accompanied by my mother and my three younger sisters, while Sumiko, the oldest, stayed in Tokyo with me (since Sumiko was already a piano major at the Ueno National Music Conservatory and I was attending Musashi by then). In Tokyo, Sumiko and I, together with two of our cousins, lived in our Magome house overseen by our grandfather and aunt Mitsu who prepared all of our meals.

At the Music Academy in Budapest, my three other younger sisters receive excellent training in cello, violin, and piano, respectively, with Futaba eventually emerging as a notable concert pianist who introduced many of Fauré's pieces to Japan. Futaba's performances, with the flutist Jean-Pierre Rampal, have also become a favorite in many quarters.

One evening in August, 1940, while in Budapest, father was listening to the 9 o'clock shortwave news program from Tokyo to catch up on events in Japan. In it, he was shocked to learn that Matsuoka had summarily dismissed some 40 Japanese diplomats from around the globe, including father himself. The next day he received the official telegram, so the family returned to Tokyo with few belongings, through Siberia on a Soviet train.

Back in Tokyo, father submitted his resignation, which for some reason was not accepted. Instead, he was asked to head the newly established training program for the Japanese Foreign Officer's Corps, which position he eventually accepted a few months later.

Other Worlds

At this point, I shall depart from following events in my father's career, and instead return to tales concerning my own experiences in Japan during these unsettled war years.

Perception-1

In the summer of 1937, despite the brewing trouble in Europe and the bucket brigades and other war-preparation exercises that were daily increasing in Japan, my parents were able to bring the family together for a month's vacation in Nagase Beach on Miura Peninsula. There, next door to our rented cottage, lived a fishing family with a 3-year-old, very healthy-looking daughter, whom my father and his friend referred to as "Yajo" (a wild girl, perhaps after a character that father and his friend had read about in a novel) because of her totally 'uncivilized' appearance.

One day this young girl came to the porch of our cottage and asked me, "Can you see where heaven is?" Pointing with her finger above the horizon, she continued, "It's right there." That night she suddenly died. Was this sheer coincidence, or had she retained some power to perceive what we in our modern life have grown up to erase from our senses?

Perception-2

Another episode took place in the early fall of 1941. I was playing tennis at Musashi on the court near our dormitory with one of my friends, Minoru Fukuda. Looking up at the slightly hazy sun, we noticed a rather striking halo, which developed into a complete, rainbow-colored ring around the sun, surrounded by yet another ring. Lying tangent to the inner ring and facing the outer ring were four symmetrically placed arcs, again with tinges of rainbow colors. The Chinese have a specific name, 'Un' (with the U pronounced as the u in you), for this rare atmospheric display. The complex Chinese character for 'Un' is made up of the character for sun over that for army, or war. Then, in December, we learned about the outbreak of the Pacific War.

Again, was this sheer coincidence, or superstition, or a convenient whitewash used by the ancient Chinese to blame Nature on man's desire for conquest in her many wars? Or, had the ancient Chinese recognized signs of Nature that flag cosmic events, which although imperceptible to us, mercilessly govern our behavior much the way we react to pheromones?

Perception-3

The following is a story I once heard from father, suggesting the implication of how a particular individual's capacities are perceived, including by one's self.

In a small country village, there lived a young boy whom everybody considered to be the town idiot. I shall call him 'Zero.' One day his school mates decided to have fun at Zero's expense, and told him "Go to the cross road at the North end of the village and wait there until a crippled fellow comes along. When you see him, tell him you want to help him and place your hand on his shoulder. Then tell him that he is now healed and see what happens."

Well, Zero did not recognize the 'cripple' that came along as one of Zero's teasing classmates. So, as instructed, Zero placed his hand on the cripple's shoulder, who, of course, straightened right up, thanked Zero for the cure, and trotted right off. Naturally Zero was puzzled by his own surprising ability, but prompted by his mean classmates, he stood at the cross road again the following week. And, again, the same scenario took place.

By the third time, Zero was convinced that he has this unusual power, so he goes to the cross road again and waits for another cripple, when along comes a real cripple. Not fazed, Zero lays his hand on the cripple's shoulder and tells the cripple that he is now cured. Whereupon much to the cripple's surprise, but not to the surprise of Zero himself, the cripple finds himself completely cured. And from that episode on, the 'town idiot' turns into a miracle worker revered by all of the town folks.

Perception-4

At the dorm at Musashi, I experienced an interesting and, even to this day, shocking discussion relating to perception. All of my fellow students insisted that the reason the rising and setting moon and sun appeared so big was because of a physical phenomenon. To prove my point, that it was not physical but psychological, I placed a small reticule in a telescope and measured the width and height of the rising and setting sun over several mornings and evenings.

What I found by using the reticule was that not only did the width remain unchanged, but that the height shrank noticeably when the sun or moon was just barely touching the horizon. But to my utter surprise, my dorm mates would not accept my observation and insisted that the large appearance of the rising and setting sun and moon was based on refraction of the rays by the Earth's atmosphere "because our teachers told us so!"

What this surprising experience makes me wonder is: Today, in the early twenty-first Century, despite modern marvels of computer tomography and nuclear magnetic resonance imaging, are not many, if not most, individuals still living in a Newtonian era of cognitive science? And when we reach the stage equivalent to Maxwell's insight into electromagnetic waves, and even to the age of quantum physics and beyond, how prudently will mankind behave and not succumb to the temptation to grasp for ultimate power? Will we by then be free not to be dictated by words 'from authority?'

War Years in Japan, Continued

Tohdai

In 1942, as students in science, we were deferred from military service and allowed to proceed to University. My heart's choice was physics or engineering, but my parent's physicist friend, Dr. Kobayashi, advised that those majors would be too difficult for me. Why not try for the Zoology Department at Tokyo Imperial University (Tohdai) since competition should be much less severe. Indeed, there was little competition; only five of us applied to the Zoology Department and all five were accepted.

When we showed up at the basement of the 2nd Science Building at our appointed time, a fellow that looked like a janitor appeared and told us to make sections of preserved frog embryos. We were then to stain and sketch the sections observed with the microscope and color the tissues according to ectoderm, endoderm, and mesoderm. It turned out that 'that janitor' was the developmental histologist, Professor Takewaki, whom we never saw again until the end of the term.

I did learn to use the series of alcohol of increasing concentration, and finally 'absolute alcohol' (dehydrated with crystals of copper sulfate), to dehydrate the embryos pickled in Bouin's solution, and then to imbibe them with grades of increasingly harder paraffin melted in a copper, paraffin oven, heated by a light bulb. The tissue in the final paraffin bath was poured into tiny paper containers to generate tiny paraffin-embedded tissue blocks. The blocks were then sliced to generate tissue sections a few microns thick using the Cambridge rocking microtome equipped with our own hand-stropped razors.

Those primitive looking but highly functional microtomes had been designed in Cambridge, England, using semi-kinematic design. I was most impressed with the kinematic design principle, based on supporting all mechanical parts by the minimum number of points to avoid over-constraining. The kinematic design achieves very high precision of movement coupled with reliably repeatable positioning.

As to making all those boring colored sketches, I tried my hand at a few but ended up fibbing by handing in sketches copied from upper classmen's work.

Another class that threw me off was plant genetics. For 2 days, Professor Shino-Toh showed slide after slide of black shadows of chromosomes until I finally gave up going to his class. It turns out that this was the famous person who had deciphered the lineage of modern rice from its ancestral, barely edible, wild form by examining its chromosomes. Little did I realize that those dark shadows of chromosomes could shed so much light on the history of agriculture.

Perhaps this also foreshadowed my poor performance in the graduate genetics class at Princeton, where Kenneth Cooper in exasperation wrote on my exam papers "Perhaps you should go back and take Genetics-1."

At Tohdai, I enjoyed some classes, such as animal physiology taught by Professor Kama-Da who had trained in A. V. Hill's laboratory in Cambridge, England. I also enjoyed the plant physiology class taught by Professor Ta-Miya, whose contributions to photosynthesis mechanisms, and for the mass culture of *Chlorella* as a food source, are well known and who happened to be the uncle of one of my friends at Musashi. In Professor Tamiya's class, we held hot debates on such topics as whether the second law of thermodynamics could not be overcome with a Maxwell's demon when one were dealing with molecular dimensions as small as those seen in the machinery of living cells.

In Professor Kamada's lab, I recall the fun of repeating A. V. Hill's thermocouple experiment for measuring osmotic pressure. By placing a tiny drop of the sample (whose osmotic pressure is unknown) on one side of a thermocouple pair and another drop of known osmotic pressure on the other side, one could compare and determine the relative osmotic pressures of the solutions by their heat of vaporization.

And then there was an alternative, if time-consuming but elegant, method for determining osmotic pressure, again of Cambridge University origin. For this alternate method, we would draw out a glass tubing into a capillary (1/16 inch or so in inside diameter) to make several 4-inch-long sections. In addition, we would prepare standard solutions (with, for example, 0.5, ca.1, and 1.5 times the expected osmotic pressure of the unknown solution). In each capillary, we would introduce micro-liter droplets of the unknown solution alternating in tandem with one of the standard solutions of known osmotic pressure. [So in each of the three capillary tubes, we would end up with a column of around 1 cm of the known, 0.5 cm of air, 1 cm of the unknown, another 0.5 cm of air, 1 cm of the known, another 0.5 cm of air, 1 cm of the unknown, another 0.5 cm of air and finally another 1 cm of the known solution.]

By leaving small air gaps in between the successive droplets in this way, we could measure the change in length of the droplets over time under the microscope. If the known and unknown solutions had the same osmotic pressure, the lengths of the droplets in that capillary would not change, since each air gap would be saturated with the same vapor pressure from the solution on both sides. But in those that were not balanced, water would evaporate from the lower osmotic pressure solution to the higher one so that their lengths would change. A graph plotting the rate of change versus tonicity of the known solutions would yield a straight line, which intersects the horizontal axis, providing the value of the osmotic pressure we were looking for.

Dream of Dividing Spirostomum

On the window sill in my room at our home in Magome, I had a bottle of pond water to which I had added a couple of short, dead branches from a grape vine. Somehow this brew turned out to be just right, because for two long years I could always find some *Spirostomum* in this bottle. But I was puzzled since I had never seen these 2-mm-long ciliated protozoa (Fig. 1.14, that look like a stretched-out *Paramecium*) undergo cell division.

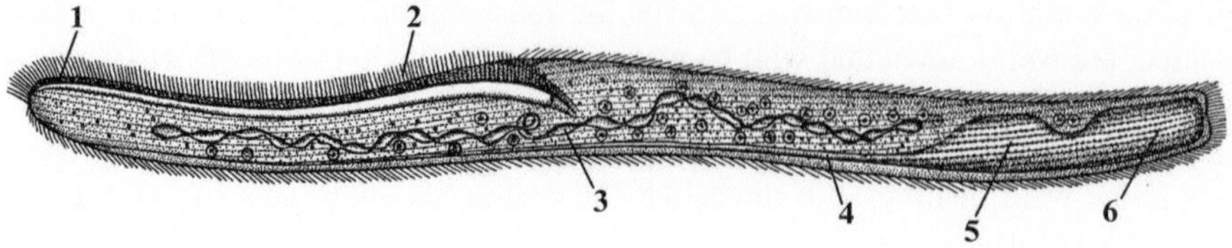

Fig. 1.14 *Spirostomum ambiguum* Ehrenberg, following Stein from Bütschli. After Karl G. Grell (1973) *Protozoology, Fig. 418,* Springer-Verlag, Berlin; and Libbie H. Hyman (1940) *The Invertebrates: Protozoa through Ctenophore, Fig. 63C*, McGraw-Hill, New York. *1* adoral ciliary membranelle; *2* peristomal cilia; *3* macronucleus; *4* canal; *5* contractile vacuole; *6* Myonemes (Courtesy of SI)

One morning I dreamed that the *Spirostomum* were finally dividing, so I took out my Reichert microscope that father had brought back for me from Germany and scanned a drop of the bottled water. Sure enough, there were *Spirostomum* swimming around, but again with no sign of their division. How in the world could they continue to be present in this bottle for so long and seemingly look perfectly happy without multiplying?

Very early the next morning, I think it must have been around 5 o'clock (a very unusual hour for a night owl like me), the puzzle was solved. I took another drop and looked under the microscope. Lo and behold, there were several *Spirostomum* at different stages of division. In those, the macronucleus had transformed into a single long sausage shape which stretched almost the whole length of the organism. In addition, halfway down the length of the long channel of the contractile vacuole a second vacuole was forming. Those *Spirostomum* had also formed a second mouth halfway down its body length. Some of them had passed these stages and started to pinch apart into two. Others were already swimming apart as two half-length ciliates.

Of course, that was a most exciting morning for me, and while making sketches of the various stages of division, I wondered whether the time of day had something to do with my finally seeing these cells divide. Well, it turned out that, after that one episode, all I had to do was to take a drop from my bottle and look at it any time of day, and there they were; some cells were always dividing. So my dream did help. But why was I not able to catch any dividing cell for two long years, until after I had seen that dream?

The fascinating changes during cell division in protists has intrigued me over the years, and that was one of the reasons that I had invited Hidemi Sato, an expert protozoologist, to join us when I chaired our Anatomy and Cytology Department at Dartmouth Medical School. And even though Hidemi kept up his ciliate culture for as long as we were at Dartmouth, I never did get around to taking advantage of his cultures. But as described later (see also *Collected Works* A-30, 31), I did have a wonderful chance to observe the eye-opening division in another unusual protist, *Barbulanympha,* many years later.

Masao and Dichrome

While I was a student at Tohdai, one of my favorite cousins was father's elder brother's (Shanghai uncle's) second son Masao Inoué. He was cheerful, open minded, and we shared an interest in science. Being several years older than me, he was drafted into the army and became a science liaison officer with industry. From one of his contacts, Mitsubishi Electric, he brought me a sheet of 'Dichrome' filter. The Dichrome sheet, which was similar to Polaroid, was my first introduction to light polarizing material. It was fabulous; it only let components of light waves vibrating in certain directions pass through. So the brightness of a scene viewed through two sheets of Dichrome changed as the orientation of one sheet was changed relative to the other. When a sheet was turned until its axis was 'crossed' with the other, no light came through.

As described a little later, this was the material I used to study the birefringence of cross-striated muscle fibers (*Collected Works* A-3). But first I used Dichrome sheets to make a projection stereopticon. For that, I made a dark box with a 100-W light bulb inside with a hole on top fitted with my grandfather's 3-inch-diameter magnifying lens. The lens, together with a 45°-tilted mirror above, acted as the projection system. On top of the lens, I placed two pieces of Dichrome, each covering the left and right half of the lens and with their polarization axes at right angles to each other.

When projected images of flowers and insects were viewed by observers wearing appropriate Dichrome glasses, out popped giant flowers and crawling insects in full color and in three-dimension pointing towards us from the screen. The Dichrome glasses were arranged so that only light coming through the left half of the projection lens would reach the left eye, and through the right half of the projection lens the light would reach the right eye (Fig. 1.15, far right).

Sadly, I could not demonstrate this stunning effect to my father, since by then he was blind in one eye after an umbrella accident caused by a hurried bus driver. What had happened was on one rainy day, when father got off the bus at his destination (in front of the Foreign Office), the driver took off without realizing that he had hooked a wing of my father's umbrella. A spoke in the other wing of the umbrella caught father's rimless eyeglasses and pushed the glass lens under the cornea of his left eye leading father to lose his left eye sight. [Without the sight from both eyes intact, one cannot tell which object is closer and which is further away, in other words one cannot achieve stereo vision.]

Just as I had managed with the stereopticon, by placing two half-shade Dichrome filters below the condenser or above the objective lens, and complementary ones above the binocular eyepiece, I was able to gain stereoscopic images of the specimen through a high-powered compound microscope ***equipped with a single objective lens*** (Fig. 1.15). For these stereoscopic viewing devices, I obtained a Japanese patent #166528 in 1944 (*Collected Works* A-2).

Years later, when I arrived in Princeton, I was surprised to find my system listed in the Bausch & Lomb Optical Company catalog. Apparently there was something

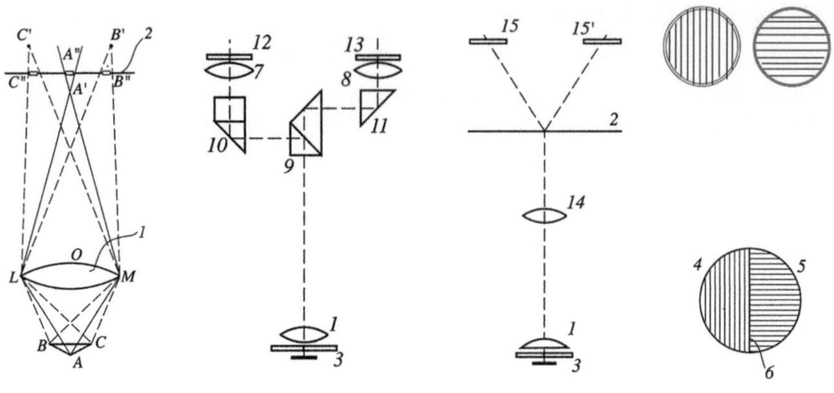

COLLECTED WORKS A-2 translated by SI from Japanese 1/6/06

PATENT NUMBER 166528 (JAPAN)
Group 3-15 Optical equipment (2, Microscopes)
Application number: 1943-11790
Application date: September 6, 1943
Patent awarded: August 18, 1944
Patent owner (Inventor): Shinya Inoué
of: 895, Magome Higashi 2-chome, Ohta-ku, Tokyo

Fig. 1.15 Dichrome Stereopticon (Inoué S, Japanese Patent #166,528: Stereoscopic apparatus, 1944. [A 2])

called an alien patents rights, which made it legal during the War to simply usurp enemy patents. However, upon my remarking to a B&L representative that I was the inventor of this patent, the item quickly disappeared from future issues of the catalog.

Spirit Over Matter

One day in 1942, the Army sent a high-ranking officer to "bring the science students of Toh-Dai up to date on the progress of War." After his remarks on the War's progress, he was asked about the impact of RADAR. [RADAR is an acronym, (first used by the U.S. Navy in 1941) for Radio Detection and Ranging. RADARs can be used to detect the location and speed of aircrafts, ships, or other metallic objects from long distances.] His response was "Nothing can overcome the strength of the human spirit. There is nothing to worry about RADAR." This shocking declaration was made to the students majoring in science shortly after the Battle of Midway!

That was the same year that I was prevented from visiting Dr. Shunshiro Kondo, who was in charge of neurosurgery at the Toh-Dai Hospital. I had gotten to know Dr. Kondo well, since he was a close friend and college soccer team mate of Katsuma Dan's. His discussions on brain legions and the loss of specific types of memory associated with the area in the brain were fascinating, and I considered him my confidant in medicine for many years. In any event, as I went down the hall towards Dr. Kondo's office, out pops a young MD in a white gown from a side door. He blocked my way and said, "You can't go in there. That's the corridor to the Ohsensei's (Great Professor's) room." No explanation from a young student, such as me would clear the way. Such was the reverence given to authority figures in Japan in those days.

An Exciting Evening

One day in 1943, Katy showed me an orange-colored book *Die Doppelbrechung*… written in 1937 by W. J. Schmidt in Giesen, Germany (Fig 1.16, far right). In the book, Katy pointed to some photographs of 4-cell-stage sea urchin eggs that Schmidt had taken through a polarizing microscope. What I saw in Schmidt's photomicrographs (Fig. 1.16, middle panel) were a single football-shaped birefringent body (with small gaps half way along their lengths) in each cell that was just about to divide.

Schmidt initially thought that those football-shaped birefringent bodies were chromosomes in the dividing cell, but 2 years later decided that they were the

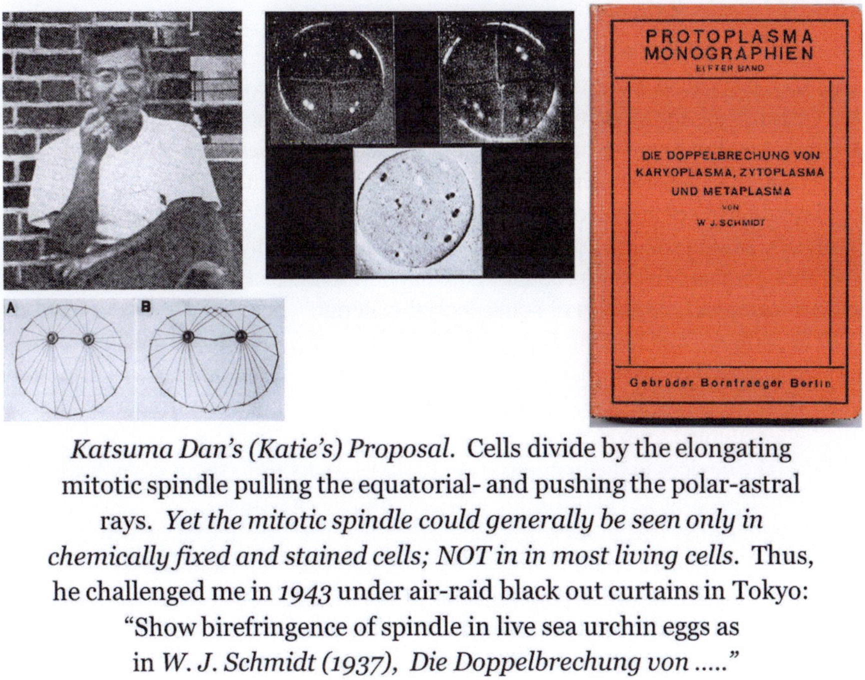

Fig. 1.16 Katy's proposal regarding cell division of sea urchin eggs (Courtesy K Dan)

spindles themselves. Those were of great interest to Katy (Fig. 1.16, upper left), because of his interest in the mechanism of cell division.

Specifically, Katy had been proposing that the elongating spindle was responsible for pushing apart two semi-rigid asters. And that by the elongating spindle applying traction and pushing forces on the fibers that made up the asters, each cell was cleaved into two daughter cells (Fig. 1.16, lower left).

A few days later, Katy invited me to his home in Kudan, Tokyo, where he and Jean had set up their temporary home away from Nagai. Katy had brought some sea urchins from Misaki in sloshing buckets of seawater and had borrowed a polarizing microscope from a friend in the Geology Department.

The excitement mounted in the small room, curtained with air raid shutters, as we tried to repeat W. J. Schmidt's observations. The eggs were collected by injecting KCl solution into the sea urchin's body cavity and then fertilized by mixing with sperm. Fifty minutes later the eggs were dividing under the polarizing microscope. But could we see the spindle-shaped birefringence? Some thought maybe yes, others definitely not. A wonderfully exciting evening; but the results were inconclusive.

The experience however, became a new challenge for me that I was to confront 4 years later in Katy's lab at Misaki after the War was over.

Frog Muscle Relay

One day, my friend Tsuruta from the Musashi days came across the campus from the Chemistry Labs to visit me in the Zoology Labs. By then, he was a star chemistry major at Toh-Dai. He had worked out the angle between the two hydrogen and oxygen atoms in water, using its infra-red Raman spectrum.

Tsuruta had come to see if I could help out with a Navy project that his mentor, Professor Mizushima had contracted. Being a top-rated physical chemist, the military thought that Mizushima could figure out how to make a sensitive relay based on electro-chemical properties of some special material.

The military was looking for some alternate means, since electronic technology in Japan those days were inadequate for making a relay that could act as a proximity fuse to trigger an explosive. The idea was to send up balloons that supported a sensitive relay coupled to explosives, to create an aerial minefield by which to deter the B-29 s. The devices in the balloons were to act as weapons to shoot down the planes that were flying far too high for the anti-aircraft guns to reach.

My suggestion was to sacrifice some frogs and use their leg muscle and associated nerve. The nerve could be connected to a simple circuit from the sensor (e.g., a dynamic microphone or dipole antenna, through a solid-state rectifier), and the muscle would be attached to the arm of a micro-switch.

I made the following for Tsuruta, which he took to Prof. Mizushima. To form a convenient base and electrical contacts for the nerve/muscle relay, I used the base that I had removed from an old 4-pronged vacuum tube. I connected wires from two

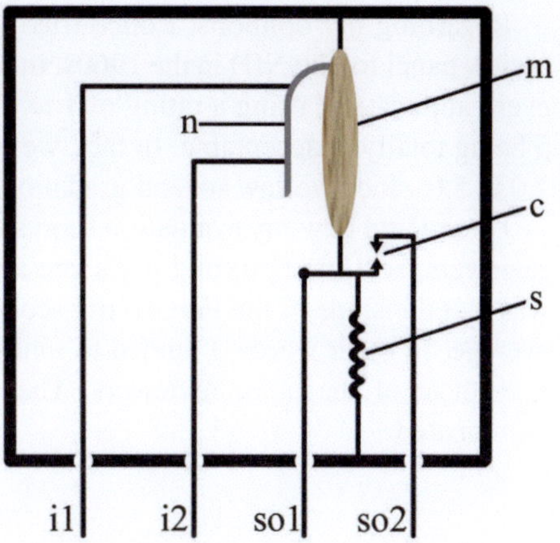

Fig. 1.17 Schematic of frog nerve/muscle relay. *m* *Gastrocnemius* muscle, *n* Sciatic nerve, *c* Electrical contact of relay output, *s* Spring, *i1*, *i2* Input to nerve from base of four-pronged vacuum tube, *so1*, *so2* Signal output from relay, connected to two other legs of the four-pronged vacuum tube base (Courtesy of SI)

feet of the vacuum-tube base to the nerve, and the other two to a hand-made micro-switch attached to the muscle. Then I covered these wired parts with an inverted test tube (that snugly fit into the vacuum-tube base) after applying to its inside a layer of filter paper wet with Ringer's solution. The moistened filter paper kept the nerve/muscle preparation in physiological shape (Fig. 1.17).

At the test field behind the Chemistry Lab, the frog relay responded perfectly – with a definite click to an extremely weak pulse of short-wave, long before the galvanometer that the technical officers had been using would even register any pulse.

I'm told Professor Mizushima won some kind of prize for this invention, but in retrospect, such a wet relay probably had little chance of working at the freezing temperatures way up in the atmosphere, tens of thousands feet high. But the story doesn't end here.

How Reliable Is My Memory?

This is where I should make a confession about my memory and strong opinions. Around 1948, I was cleaning up the bookshelf in my study at our Magome home, preparing for my trip to Princeton. It was then that I came across a composition that I had written as a sophomore in high school.

The composition turned out to be about Marconi's invention of the telegraph. What should I find in the composition but, "In his early attempts to confirm that the wireless signals were actually being transmitted, Marconi watched the synchronous twitching of a frog muscle whose nerve was connected to the receiving antenna."

That was written in my own hand half a decade preceding my 'invention' of the frog muscle relay! In high school, I had obviously read about Marconi's experience, written about it, and then had completely forgotten about that. And now I am writing about many things that happened over 80 years ago. I'm trying hard to be truthful, but how well can my memory be trusted?

Regarding my opinions, I once tried a test on myself when serving on a grant review panel for the NIH in the 1960s. In those days, the panel members each graded every application, using a rating of 1 to 5, with 1 being absolutely outstanding and 5 being totally unacceptable. In fact, we expressed our views by using ratings from 1.0 to 5.0, since we saw several gradations between 1 and 2, 2 and 3, and so forth.

I wondered how my evaluation scores compared with the average of all ten or so reviewers, so I tried plotting my score against the average score. What turned out was that the slope of the line for my scores was noticeably steeper than that for the average. In other words, I tended to rank high scores higher, and low scores lower than most of the other reviewers. That told me that I must be rather strongly opinionated.

Dissertation at Tohdai

For my graduation dissertation at Tohdai, my assignment was to follow up with the work of Hans Spemann. He had reported that a lens excised from the eye of a newt would not only grow back, but do so by forming a new lens growing out from the upper edge of the iris of the same eye. Not only was the regeneration of a functioning lens of interest in itself, Spemann had found that the iris from which the lens regrew was of mesodermal origin, rather than a product of the epidermis (as was the lens originally formed during normal development). In other words, lens cells could be induced to be formed from a tissue layer (mesoderm) differing from the tissue layer (ectoderm) which initially gave rise to the lens.

So I first learned to extract a lens from the eye of a newt, then use a fixative which would not harden the lens, but (unlike many other fixatives) leave it soft enough to be cut with a microtome knife. Then I sacrificed newts with an enucleated eye at 1 week intervals. After removing the eye with the regenerating lens, I made sections of the eye and examined with a microscope the stained sections showing the regenerating lens. [Not the kind of biology I was very fond of, but anyhow I was able to confirm Spemann's findings which allowed me to graduate from the Zoology Department.]

After Graduation

Thermocouples

When I was about to graduate from Tohdai, in 1944, Tsuruta appeared again and suggested I join him in a military research project. The project was to develop sensitive infra-red detectors, so that they could be used instead of humans (i.e., Kamikaze pilots). In the absence of such detectors, soldiers who were guiding planes were being sacrificed as bombs, or those steering the miniature submarines were sacrificed as torpedoes.

Not only would the sensitive infra-red detectors replace human sensors and prevent human sacrifices, but the project would also keep us from being called and sent to the hopeless front.

By then in Japan, there were few scientists who were not involved in some form of War effort, except for Katy who stuck with his studies on cell division. As a Japanese youngster, the Kenpeitai (the Japanese military police) had absolute power over our lives, and I had not even known of the concept of conscientious objection until I arrived at Princeton.

After studying a pirated version of John Strong's *Procedures in Experimental Physics*, Tsuruta and I made parts for sensitive thermocouples by drawing up molten Tellurium into glass capillaries. By placing one end of the glass capillary in our mouths, we sucked molten Tellurium up into the capillary, not realizing how toxic its vapor was.

After cooling, we took the Tellurium rod and dissolved away the glass tubing with hydrofluoric acid. To one end of the Tellurium rod, we welded on a thin ribbon of soot-covered Constantan foil, thus generating the sensory junction for the sensitive thermocouple. Then, we placed the part of the soot-coated Constantan foil welded to the Tellurium rod exactly at the focus of a parabolic mirror. Using such a device connected to a ribbon galvanometer, we could detect the heat generated by ourselves on cold days standing 100 feet away.

We could also detect from the Yokosuka hilltops the newly built aircraft carrier in Tokyo Bay. How we got away without the Kenpeitai catching such suspicious looking youngsters, pointing an infra-red detector on a battleship, I will never know.

Imperial Guard Supply Corp

The military was pretty satisfied with our efforts, and I was not called into active service until the late spring of 1945, 4 month before the end of the War. Since I had declined the offer to become a lieutenant to stay with the project, I was drafted as a buck private in the Imperial Guard Supply Corp. (Remember, I was a zoology major; zoologists are supposed to know all about horses!).

But what a sad Army it was by then. Even though we were supposed to be in the prestigious group guarding the Imperial City, we had no leather shoes nor belts and sabers, so were given slips of paper stating why we were not in proper uniform in case the Kenpeitai (the dreaded Japanese military police) stopped us. Among those called up at the same time as myself were three active epileptics who had several fits a day; in other words, we were truly the dregs of the barrel.

Our sergeants had fits watching us chew our food. They were hand tied by the new regulations, since some thirty of those inducted half-a-year earlier had died of malnutrition as they were commanded to swallow the poorly-cooked millet. Until then the Japanese Army had prided itself in the speedy meals, which was actually possible with soft cooked rice, but rice was no longer available even for the Imperial Guard Supply Corp.

Soldiers on leave to visit their farm families were told to bring back a 2-l bottle of pine oil distilled from old pine roots. (Before the War, pine roots were left in the ground after the tree was felled for lumber). "We can then move the trucks which can bring us some rice" was the reason given (because gasoline and diesel oil which would otherwise be used to fuel the trucks were not available, even for the Imperial Guard Supply troops).

In a few weeks after I was drafted, came the great carpet bombing of Tokyo. By then, we had only been introduced to the horses twice, but had to lead the panicked creatures away from the burning stables. With the horses safe, we dashed out to the street to help quell the burning houses, and those whose clothing were burning with napalm. The scattering bomblets of napalm sounded like sand being sprinkled on water, but strangely I don't remember being scared of them raining down around us.

With our barracks in ashes, we were sent to different troops, in my case to Kohfu. As it turned out, that was not very far from where my parents and sisters were staying. As liaison to the German Embassy, father had been appointed Ambassador to Germany and stationed by Lake Kawaguchi on the other side of Mt Fuji.

After seeing the secret message I sent from a drug store in town (implying which district I had been sent to, after the carpet bombing of Tokyo), somehow my parents and the research team located me, so I was back once again with my family and Tsuruta and company. I had been to the drug store several times, since in the absence of proper medication, I was helping those in the troops suffering from malnutrition and intestinal disorders by procuring and distributing vitamins and active-charcoal pills. My own GI track was also in pretty sad shape by then, and I contributed little in the last month to the research team that was housed in a nearby school building.

Somehow, my parents and sisters and I were now all living together in a room rented in a local doctor's house in Hinoharu, outside of Kohfu. To the doctor's wife, my westernized sisters, in particular, were a real curiosity, but at least we were able to sustain ourselves for a while living in this farm community. It was there that we heard the Imperial Edict declaring the surrender of Japan to the Allied Forces mentioned in the PROLOGUE: *The Last One to Go.*

How Muscle Contracts

As part of my graduate work at Tohdai, in 1947 I joined the group in Professor Kamada's lab where people were isolating single, functional, frog muscle fibers. Under a dissecting microscope, we very carefully removed excess fibers from an excised frog muscle using sharp iridectomy scissors and fine, sharpened forceps. The muscle itself was suspended by its tendons between two tiny hooks in a chamber filled with Ringer's solution.

When the operation was successful, a single fiber, less than a hundred microns in diameter, remained intact along its full length from one tendon to the other. The isolated fiber was judged to be intact when the fiber responded to a DC stimulus with a single twitch and no clot appeared on the fiber.

While most of the folks in the Kamada lab were interested in the electrical properties of such fibers, my interest was in looking for the mechanism of contraction of the isolated muscle fiber upon stimulation.

Being skeletal muscle, these fibers had beautiful cross striations repeating every two to three microns. As had been known for many years, each cross striation is made up of a birefringent (or anisotropic) 'A-band' sandwiched by an 'I-band' which shows very low birefringence (hence the name Isotropic-band, in contrast to the Anisotropic-band).

In the field of the microscope made dark by placing a sheet of polarizing filter (the polarizer) below the condenser, and another one (the analyzer) placed above the objective lens and with its transmission axis crossed with the polarizer's, the birefringent A-bands shine brightly alternating with the dark, 'isotropic' I-bands on a dark field. I was curious to know what happens to these bands as the muscle contracts.

So I set up a microscope to view the image of a muscle fiber between crossed polarizers, with the length of the fiber oriented at 45° to the polarizer axes. In addition, I placed a quartz wedge in front of the polarizer (located some distance from the condenser), so that the condenser would project an image of the quartz wedge superimposed on the image of the muscle fiber.

By orienting the quartz wedge so that its length runs parallel to the length of the muscle fiber, the birefringence of the muscle (A-bands) is 'compensated' by the birefringence of the quartz wedge. [Note: A quartz wedge is manufactured in such a way that the slow axis of its birefringence lies perpendicular to the length of the wedge, a direction opposite to the slow axis direction of muscle (and many other types of) fibers.] Also, the quartz wedge is made long enough so that its birefringence retardation equals 1, 2, 3, etc., times the wavelength of the (monochromatic) illumination used. Those locations (thicknesses) of the wedge can be identified by the dark fringes that run across the width of the wedge.

Thus, with monochromatic illumination, the dark fringes, which run across the quartz wedge, are displaced proportionately to the thickness of the muscle fiber (which reflects the birefringence retardation of that portion of the muscle fiber). The result is a parallel series of dark fringes in the shape of semi-ellipses that reflect the cross-sectional shape of the fiber (Fig. 1.18).

In addition, I found that if I overlaid part of the quartz wedge with a sheet of mica of a particular thickness, I gained semi-elliptical fringes which were shifted in the opposite direction compared to the fringes seen without the mica sheet. [It turns out that this particular mica sheet was introducing exactly half a wavelength of retardation, hence a shift in the location of the compensated dark fringe by 180°; in other words to a symmetrically opposite (thinner) side of the quartz wedge.]

Thus, by overlaying the mica sheet so that its edge lay across one of the dark fringe regions of the quartz wedge, I gained a full elliptical fringe which reflected the birefringence (retardation) of the full cross section of the muscle.

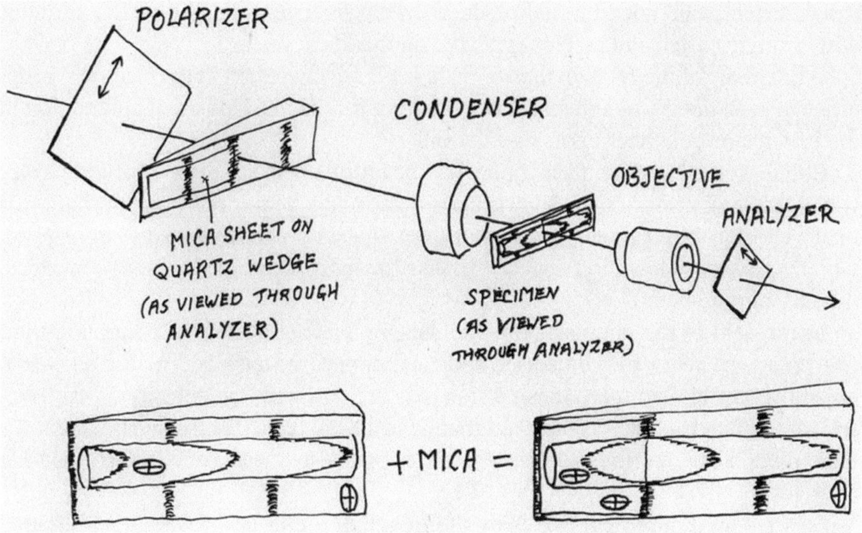

Fig. 1.18 Muscle fiber viewed between crossed polarizers, superimposed on image of quartz wedge. The wedge is partially covered with a thin sheet of mica and viewed in monochromatic light. (From *Collected Works* A-3.) (Inoue S, Birefringence versus length of resting and contracting single muscle fiber. Unpublished studies 1947: Collected Works of Shinya Inoué, World Scientific Publishing, 2008)

The point of setting up such a contraption was so that I could cut out photographs of the ellipse and weigh them to measure the changes in their area (namely, the integrated birefringence of the muscle) during contraction.

What I found was that the cross-sectional birefringence of the fiber was only a function of the fiber length (decreasing as the square root of the length) and **not affected by whether the fiber was stimulated and contracting or not**. In other words, the birefringence per unit thickness of the fibers did not change, whether the fiber was passively being stretched or was shortening by electrically stimulated contraction (Fig. 1.19).

That meant that the contraction of striated muscle seemed quite different from that in rubber bands, which, as we saw earlier, is brought about by folding of polymer chains and results in a drop of birefringence associated with its contraction.

Well, Professor Kamada was not interested in my results and even told me that one could not learn about muscle contraction by studying skeletal muscle, since it has all those strange cross striations. In fact, he continued, "You should study isolated, single smooth muscle cells if you want to find out how muscle contracts."

Alas, it was some 10 years later that Hugh Huxley and Jean Hanson, and Andrew Huxley and Rolf Niedergerke, reported that muscle contracts by the mutual sliding of the protein filaments that make up the A- and I-bands, rather than by folding of filaments. And they had developed the sliding filament model by studying the optical properties of cross-striated muscle (fibrils and fibers).

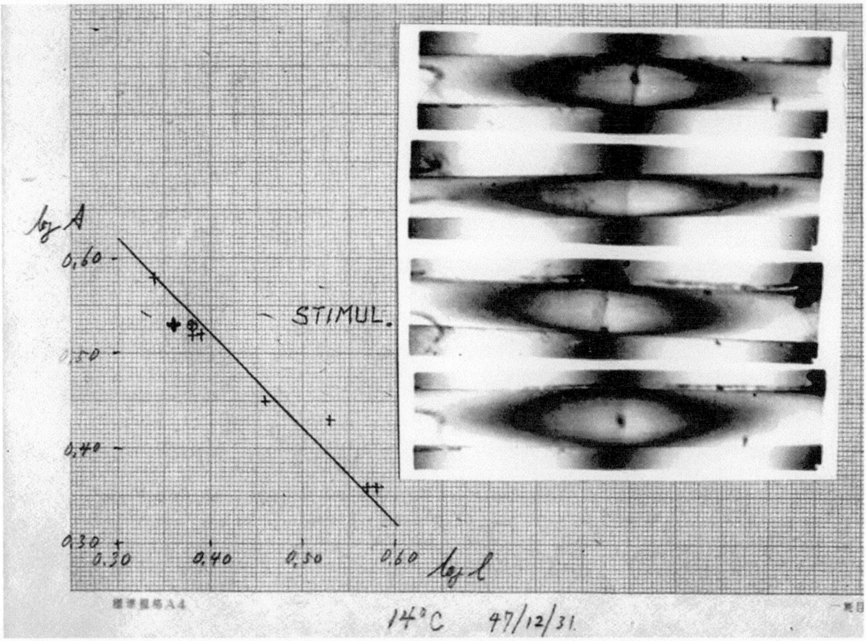

Fig. 1.19 *Left*: Log/log plot of birefringence (retardance) versus length of a single cross-striated muscle fiber. *Right*: Photographs of the single muscle fiber stretched to different lengths, without stimulation (*top two*), and stimulated (*bottom two images*). (From *Collected Works* A-3.) (Inoue S, Birefringence versus length of resting and contracting single muscle fiber. Unpublished studies 1947: Collected Works of Shinya Inoué, World Scientific Publishing, 2008)

In retrospect, Professor Kamada must not have been interested in welcoming *'unusual cells,' which are special gifts of Nature that help us reveal her mysteries.*

[These results of my experiments and the accompanying illustrations never appeared in public until they were published as Article 3 in *Collected Works of Shinya Inoué: Microscopes, Living Cells, and Dynamic Molecules* published in 2008 by World Scientific Publishers.]

Soon after these events, Professor Kamada succumbed to terminal cancer and passed away in 1947, so I decided to work with Katsuma Dan, who by that time had been appointed an Instructor in the Zoology Department at Tohdai.

For my work space, the Chairman of Zoology, Dr. Y. Okada, assigned me to Dr. Sohmei Abbe's lab. This laboratory in the basement of the science building was packed with 3-feet-tall specimen jars containing formalin-pickled fish since Dr. Abbe was an ichthyologist.

But Dr. Abbe kindly moved some of the jars out into the hallway and had a desk brought into his room as work space for me. There I finished analyzing the muscle data.

Many years later, Katy told me that placing me in that room with pickled fish was the Chairman's way of punishing Katy for daring, as a young instructor, to take on a graduate student to supervise!

Chapter 2
The War Ends

Immediate Post-War Years

Nagai then Misaki

Soon after August 15, 1945, my family and I headed back from the farm house retreat in Hinoharu to our home in Magome. Fortunately our home had been protected by the bucket brigade and escaped becoming another casualty of the extensive Tokyo fire bombings. As we came out from the Ohmori train station, American GIs fumigated us with DDT powder, purging us of the fleas and lice that were ever present in the crowded barracks and farm communities.

Of the events during the first few months thereafter, I have little recollection, except for seeing my parents going off to the countryside to barter what little they still possessed for sweet potatoes. For months, there was nothing much else to eat until small allotments of rice began to be rationed again. I was apparently not well enough to help my parents haul those heavy sacks of sweet potatoes.

Shortly after we arrived back in Magome, the wooden roofs of the Ohmori train station and the public phone booth in front of the station, which were all burned out during the air raids, were repaired by patchwork. Still the coin system for the public phone did not work for some time. Instead, there was a wooden box with an open top next to the phone; you simply deposited a ten cent coin into the box and made your call. What surprised me was that this public honor system worked, despite the extremely hard financial time everyone was having.

In the meanwhile, Jean Dan and her five children, who were sheltered in the Hakone Mountains, returned to their home in Nagai to be rejoined with Katy. Soon Jean was running around, collecting surplus material from the occupation forces to distribute to the neighboring farmers, helping them set up a cooperative in which the wastes of war, such as the shells of the napalm bombs, were literally converted into plows. Jean even approached the U. S. Army Headquarters and successfully negotiated with them to use only half of the local airfield to build the officer's quarters.

S. Inoué, *Pathways of a Cell Biologist*, Springer Biographies,
DOI 10.1007/978-981-10-0947-1_2

Thus, she was able to get the other half of the field returned to the farmers of Nagai. The whole farmland had been usurped and converted to an airfield by the Japanese Air Force during the War.

Jean was thrilled to be helpful to her own country's occupation forces, as well as to her Japanese neighbors. In addition, the Dan's did not lack for food as the grateful farmers and fishermen brought produce to thank her for her deeds as well as for the practical lessons in democracy. When the farmers offered to do something further, she suggested they set up a fellowship for a young Japanese student. I was the recipient of this gift to Jean.

For several months, the Dans nourished me back to strength at their home in Nagai. I watched Katy prune the ancient plum trees, whose white blossoms attracted the honey bees from the hives in Katy's back yard (Fig. 2.1).

Breakfast was corn flakes sweetened with the fresh honey, and the donuts that Jean brought back from the PX, all of which were luxuries beyond belief. I still recall the walks between the fresh green rice paddies with the Dans' second son Harumi, probably 5 or 6 years old at the time, who shared my interest in watching the frogs and insects. As an adult, Harumi immigrated to Brazil and started a fish farm. Many years later Jean, who seldom complains, remarked to me that I was to blame (by my example) for Harumi's leaving Japan. Sorry Jean. But now back to 1947.

Fig. 2.1 Katy relaxing under plum trees in his yard in Nagai (Courtesy Dan family)

Experiments in Optical Design

Finally, back at the lab in Misaki, we could continue what we had left off at Kudan that memorable evening during the air raid blackout. Using a mercury arc lamp that I picked up from a surplus store to gain enough brightness, a Glan-Thompson prism donated by Professor Koana of the Physics Department of Toh-Dai to generate pure polarized light, and a Dichrome analyzer attached to the draw tube of Katy's Zeiss microscope, I assembled what I thought should be a good polarizing microscope. Katy chose the glass-clear eggs of the sand dollar, *Clypeaster japonica*, and we waited for a little over half an hour after they were fertilized. This time, the birefringent spindles started appearing, some white and the others black on a dark grey background just as Schmidt showed in his 1937 monograph, and then they would disappear just before the cells started to cleave.

Tantalized by these dynamic scenes, I wanted to follow the changes further as the egg cells developed, but felt that the scope I had put together was not quite sensitive enough. Peering at the back of the objective lens through the microscope body (with the eyepiece removed), I found that the objective lens was not as dark as expected between crossed polarizers. So I convinced Katy to let me try to improve the system so that we could see the spindles better.

Adjusting the orientation of the polarizer and analyzer didn't help darken the light coming through the lenses any further. So, suspecting that the objective lens may be strained and that it was introducing some undesirable birefringence, I tried turning the lens in its mount. To my delight, the back of the lens and the microscope field became much darker, so that is where I left the lens. But now, we could not see the birefringent spindles. They had completely disappeared, no matter how many new eggs we tried. "I told you to leave well enough alone," was Katy's remark, as red-faced I kept on trying to figure out what went wrong.

It took me nearly a month, but I finally figured out that turning the lens to gain maximum extinction had effectively eliminated an essential optical component. What was needed, I found out the hard way, was the presence of a compensator placed between the crossed polars (also known as polarizers) in addition to the specimen. Without a compensator, a specimen with as low a birefringence as the spindle in living cells would not generate detectable contrast above the dark background between the crossed polars.

When a compensator possessing a very weak birefringence is introduced between the crossed polars, the extinction drops somewhat and the microscope field becomes a dark grey instead of nearly black. However, the part of the specimen whose birefringence axes are in the same orientation as the compensator's axes becomes brighter, while the part with opposite orientations becomes darker than the background. In other words, an appropriate compensator dramatically raises the sensitivity for detecting very weak birefringence with a polarizing microscope; the image contrast is now raised to a range detectable by the human eye. [Note: the birefringence (retardation) of the spindle is very low–only a few nanometers, or one hundred times lower than the wavelength of visible light.]

In addition to vastly improving the sensitivity, the compensator allowed us to determine the sign of birefringence of the specimen, i.e., which way the molecules that introduced the birefringence were aligned. Furthermore, the orientation of the compensator that completely darkened a specimen region allowed us to determine its retardation, or show us how many molecules were lined up in that specimen region.

Of course, the molecules themselves are much too small to be resolved with the light microscope. So a polarizing microscope, equipped with polars providing high extinction, a bright light source, and an appropriate compensator, allows us to probe structures way below the resolution limit of the microscope and to determine the alignment and concentration of birefringent molecules. This can be done without killing or staining the cell, in other words, non-destructively. Many molecules in living cells turn out to be lined up in characteristic ways; they also change dynamically according to the functional states of the cell. Thus, a sensitive polarizing microscope becomes an exceptionally powerful tool for non-destructively studying dynamic molecular changes, directly inside living cells.

In retrospect, I found that the need to use a weak compensator was already known to W. J. Schmidt and even described in his monograph that reported on the spindle birefringence. But with my poor ability to read German, I hadn't even thought of looking through the first part of the monograph. So it took me a whole month to rediscover what Schmidt had already known. On the other hand, who could ever forget a lesson learned such a hard way!

In any event, by making sure that the room temperature was not too high, and that the seawater surrounding the cells did not evaporate and become too hypertonic or squash the cells, in other words under conditions that kept the cells happy and dividing, Katy and I were able to follow the changes in the spindle and aster birefringence before and during several cleavage cycles in the *Clypeaster* eggs and in eggs of a jellyfish, *Spirocodon* (Fig. 2.2). We published these results in the *Journal of Morphology* after I arrived in Princeton (*Collected Works* A-5).

Interestingly, just about the same time, Michael Swann and Murdoch Mitchison, then in Cambridge, England, had also shown how to improve the sensitivity of the polarizing microscope, introduced the definition of 'Extinction Factor,' and described the changes in birefringence of the spindle in dividing eggs of a European sea urchin, *Psammechinus miliaris.* Their observations were very similar to ours although our interpretations were not quite the same.

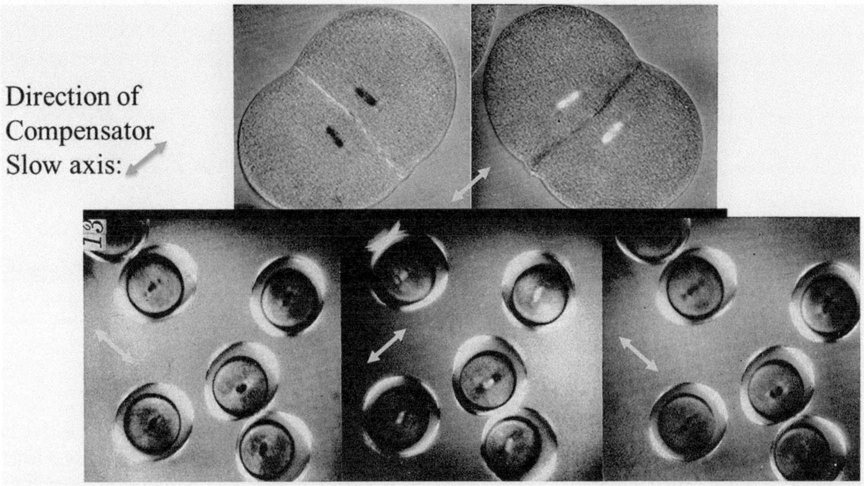

Fig. 2.2 Birefringent spindles and asters. These dividing cells were observed at Misaki using the 'Shinya-scope' in 1948. Eggs of jelly fish (*Spirocodon sp., top*) and sea urchin (*Clypeaster japonica, bottom*) show positively birefringent spindles and asters. Sea urchin eggs (but not jelly fish eggs) are surrounded by positively birefringent fertilization envelope, and its early spindle poles by a negatively birefringent 'streak' (*Collected Works* A-5) (Inoué S, Dan K, Birefringence of the dividing cell, J Morphol 89 (3): 423–456, 1951. [A 5])

Encounter with Emperor Showa

Kayo and the Sea Urchin Spicules

While I was fiddling with the polarizing microscope, two other budding scientists also frequented Katy's lab in Misaki. One was Kayo Okazaki, who had graduated from Nara Women Teacher's College and was interested in the developmental fate of cells derived from different parts of the egg. The other was Yoshiyuki Endo, who had forfeited his position succeeding the family hardware store and instead given the business to his younger brother so that he himself could follow his interest in biology, specifically the immediate reaction of the egg to fertilization by sperm.

Kayo was showing how whole embryos, and even young metamorphosed adults, could be developed from individual cells isolated from 16-cell-stage sea urchin eggs. The development of the antler-shaped skeletal spicule (the pair of bones supporting the arms) of the sea urchin's pluteus stage larvae also attracted her interest.

Again, it was W. J. Schmidt who had shown earlier that each of these pair of skeletal spicules was a 'biocrystal.' Each is essentially built with its molecules arranged as in a single crystal of calcite, although its shape has no resemblance to calcite crystals, hence the term biocrystal. Just like a tiny crystal of calcite, the spicule shines brightly under the microscope between crossed polars.

As the pluteus embryo swims and tumbles around, each spicule suddenly loses its brightness as its optical axes line up with the transmission axes of the polars, just as does a single crystal of pure calcite (Figs. 2.3a, 2.3b and 2.3c).

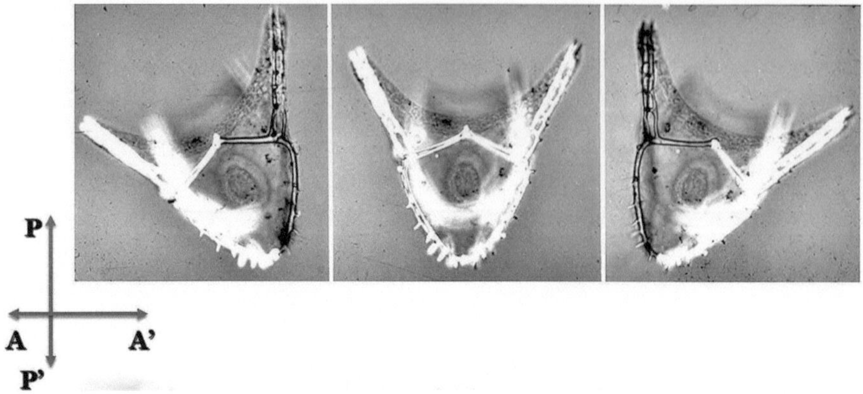

Fig. 2.3a Birefringence of pluteus spicule. Observed between crossed polarizers, the left and right skeletal spicules (of the pluteus larva of a sea urchin, *Lytechinus variegatus*) each extinguishes completely when oriented parallel to the polarizer or analyzer, reflecting its uniform crystalline nature (*Collected Works* A-29) (Courtesy of SI)

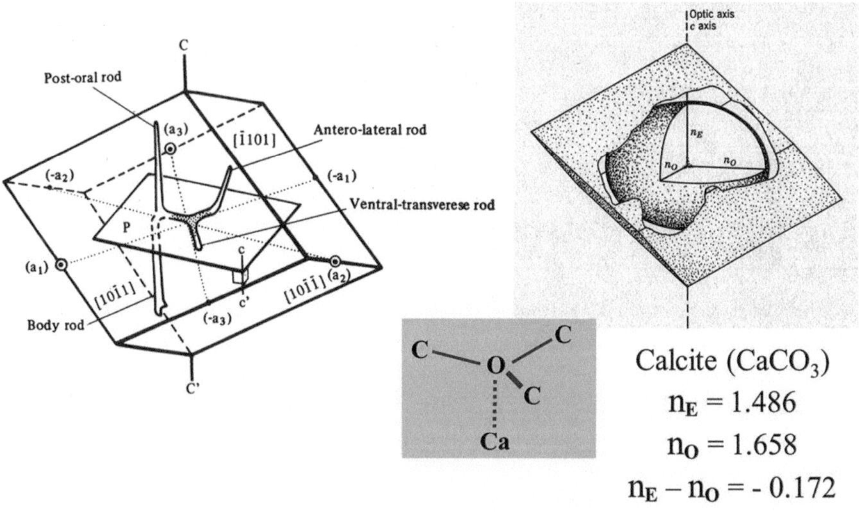

Fig. 2.3b *Left*: Optic axis of skeletal spicule, a biocrystal. The spicule (>99 % $CaCO_3$) is shown with its optic axis (CC′) aligned with that of a calcite crystal. *Right*: Calcite shown together with its uniaxial index ellipsoid (Okazaki K, Inoué S, Crystal property of the larval sea urchin spicule, Dev Growth Differ 18 (4): 413–434, 1976. [A 29])

Kayo was interested in how the few embryonic ("mesenchyme") cells, which wander into the hollow ball of cells of the sea urchin (in the gastrula stage) embryo, can form these calcite spicules patterned into species- specific shapes (Fig. 2.4). Since the polarizing microscope is a sensitive detector of the birefringence and axes of even the smallest crystals of calcite [because calcite has a very high 'coefficient of birefringence' which is equal to 'retardance per unit thickness'] it was natural that Kayo and I would work together on this project.

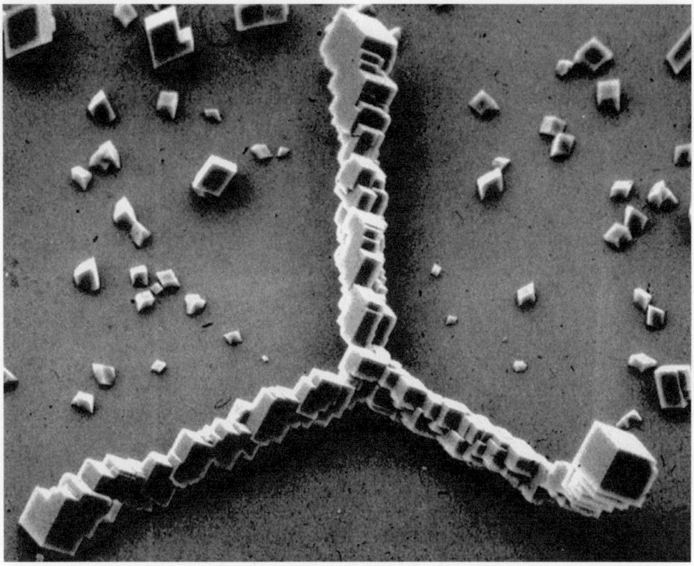

Fig. 2.3c Epitaxial decoration of isolated triradiate spicule. The spicule, isolated by immersion in a solution of $CaCl_2 + NaHCO_3$, has amplified the calcite lattice of the biocrystal. Scanning EM by Kent McDonald (*Collected Work* A-29) (Okazaki K, Inoué S, Crystal property of the larval sea urchin spicule, Dev Growth Differ 18 (4): 413–434, 1976. [A 29])

Designing Shinya-Scope 1

Then, one day Katy told us that Emperor would be visiting the Lab and that I should show him the swimming plutei with their spicules sparkling between crossed polars. Well, up until that point, I had placed the mercury arc lamp, Glan-Thompson polarizer, and the modified horizontal microscope stand each on a pile of books to align their height. With the Emperor's coming, that seemed too precarious an arrangement for the demonstration, so I went out to the woods surrounding the Laboratory in search of a suitable stand. I came across a piece of steel about two and a half feet long, an inch thick, and 4 inches wide. It was apparently a cast-away machine gun base and had a two-foot stretch whose middle was slotted, just ideal for fastening and securing the microscope parts.

Katy had no objection to my bringing in this odd piece to his lab. So after cleaning off the dirt, I tied the several microscope parts to this stand with string, using V-shaped wooden blocks to adjust the height of the lamp house and polarizer (Fig. 2.5).

This was called the Shinya-scope by Katy and Kayo after I left for the States. This became the first of the many Shinya-scopes that I would build or design in my life time.

Explaining the blinking spicules in the swimming pluteus seen through this microscope turned out to be my first encounter with Emperor Showa. He seemed perfectly at ease and quite interested looking through the jerry-built microscope and

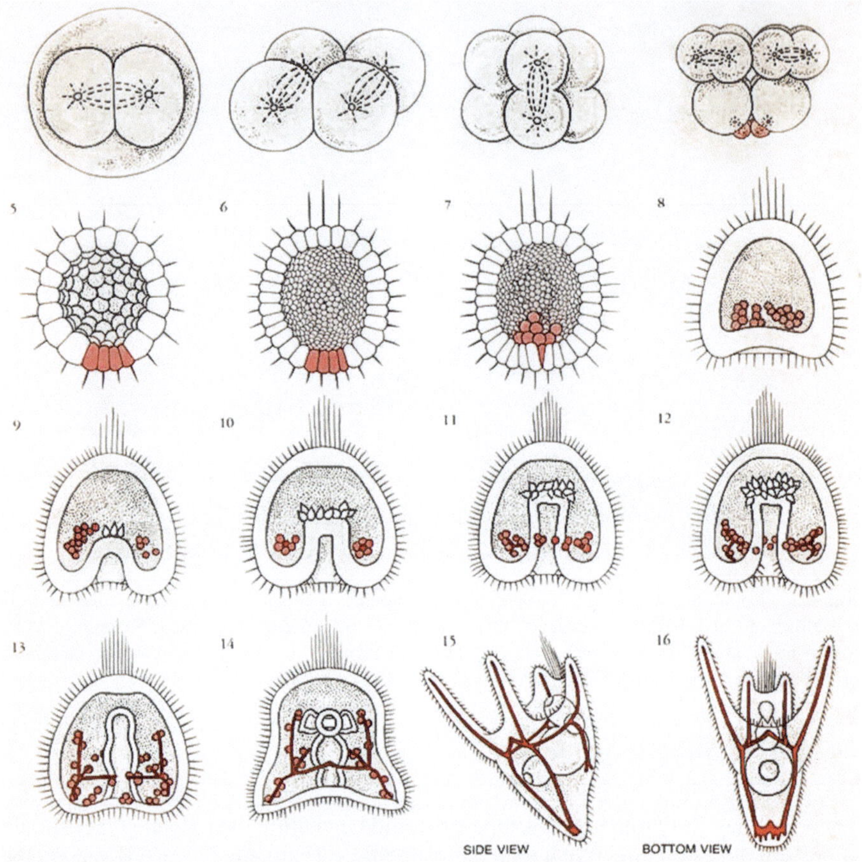

Fig. 2.4 Development of sea urchin embryo from first cleavage division to pluteus stage. Micromeres (shown in *dark grey red*), formed at the vegetal pole of the embryo at fourth cleavage, multiply and enter the blastocoel as 'primary mesenchyme cells.' The syncytial mesenchyme cells generate and pattern the biocrystalline spicules (*Collected Work* A-29) (Courtesy SI)

hearing explanations of the scene. After all, he was an accomplished biologist who had already published several volumes on the nudibranchs and polyps of Sagami Bay (not in his own name but as publications from the Imperial Laboratories, since as deity, he was supposed to be above such activities). The jolly Empress seemed to especially enjoy probing through the tide pools, and the Crown Prince (the current Emperor Akihito, now an accomplished ichthyologist, but at that time still a youngster in high school) was fascinated by the electric rays. After the presentation, I was pleased to be complimented by Katy that my speech was appropriately polite, yet free of the expressions that kowtowed to the Emperor.

While he was relaxed in the lab and during beach combing, it was a surprise to see how ill at ease the Emperor was when the Imperial party met the local entourage that had gathered to greet him and his family at the Lab entrance. Perhaps that

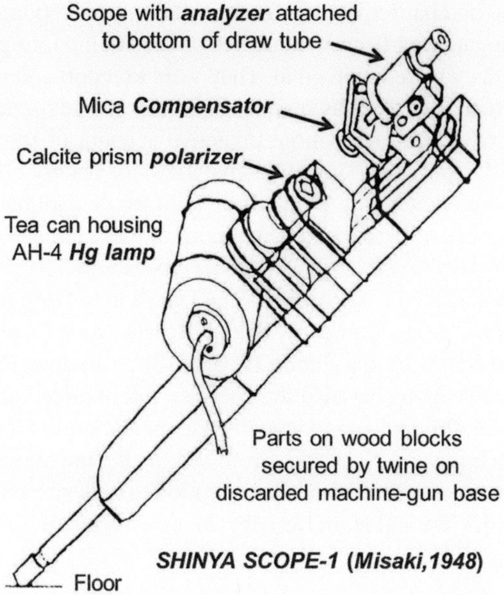

Fig. 2.5 Shinya-scope 1. Two years after the War, I re-joined Katy at the Misaki Marine Biological Station and built this polarized light microscope using available parts (Appendix I in *Collected Works*)

should not have been too surprising, since he had not long ago descended from being considered as God to the rank of humans. [Until he declared himself otherwise (at the end of WW-II), most Japanese really thought of the Emperor as being a living God].

Here was a leader who was bewildered in the presence of the folks, who formerly would have been willing to fight or die "in the name of the Emperor." Yet another facet of this person was portrayed by General Douglas McArthur, who upon his first visit as Supreme Commander of the Allied Forces was said "to have been most impressed by this gentleman, who, instead of pleading for his own life or position, took full blame for Japan's war and deeds" (Inoue K. 1991). This encounter must have had much to do with the General's insight to retain the Emperor as figurehead for post-war Japan, rather than consider him as a war criminal as many had urged.

GHQ-3

A detailed record of the events during and preceding the Pacific War was apparently collected by those working in MacArthur's SCAP GHQ-3. My father, who was summoned to participate because of his position in the Japanese Foreign Office, told me how those who had participated in foreign affairs, or as reporters or artists in the battle fields, were gathered from both shores of the Pacific into groups, depending on which conflict they were involved in. They were to report and portray their experiences in the particular conflict as seen through their own experiences, after which historians would try to extract a more objective account of the battles and other events. One would expect that such objective views of history could be invaluable for helping to circumvent future political or military misjudgments, but for some reason, I have not been able to locate any record of these collections, if they in fact do exist.

One video-tape narrative, "Hiroshima: The Decision to Drop the Bomb," assembled by Asahi Broadcasting Company, British Broadcasting Company, and distributed in the United States by the Public Broadcasting Systems, shows not only the horrors of modern weapons but also the complex international and national events that led up to the fateful decision to actually use the A-bombs (see Google: Atomic bombings of Hiroshima and Nagasaki. From Wikipedia, the free encyclopedia). We should insist that every person, who is in a position to make decisions that affect the future of this world, view and learn from the documentation of this harrowing event.

Jean Dan

Jean Dan's Homecoming and the Phase Contrast Microscope

In 1947, Jean Dan had her first trip back to the States after World War II (Fig. 2.6). She was greeted warmly by her sister Peggy Chittick and her family in Milford, Connecticut, and of course, by her many friends in Woods Hole. They must have had exciting exchanges about their wartime and post-war experiences, especially of Jean's in Japan, how the children and Katy had managed, and about the research that had gone on at Misaki. Through her letters, we learned how Katy's message '*The Last One to Go*' had made *Time Magazine*, how Jean and Katy's classmates from their Pennsylvania and Woods Hole days were faring, and about the new kind of microscope that she was bringing back as a present to Katy. The surprise gift was called a phase contrast microscope (Fig. 2.7a), which Katy insisted was not a schlieren microscope as I had thought. How Katy knew so definitely, and so correctly, is still a mystery to me.

The phase contrast microscope turned out to involve a totally new optical principle developed by Fritz Zernicke in Holland. He had shown that light diffracted from a minute transparent region of the specimen was a quarter-of-a-wavelength out of phase with the un-diffracted light. By taking advantage of this difference (which ordinarily is not detectable to our eye) and by increasing this difference by another quarter-of-a-wavelength so that the half-wave, phase-shifted diffracted light would destructively interfere with the un-diffracted light (taking advantage of the fact that diffracted and un-diffracted light travel through different parts of the objective lens),

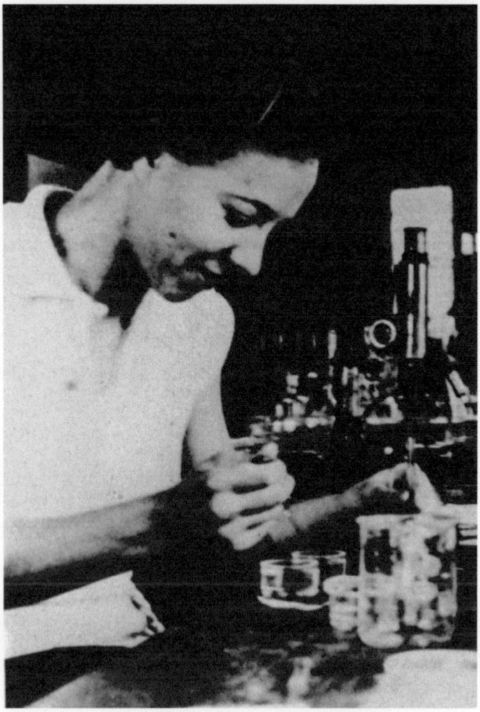

Fig. 2.6 Jean Clark Dan (Courtesy Dan family)

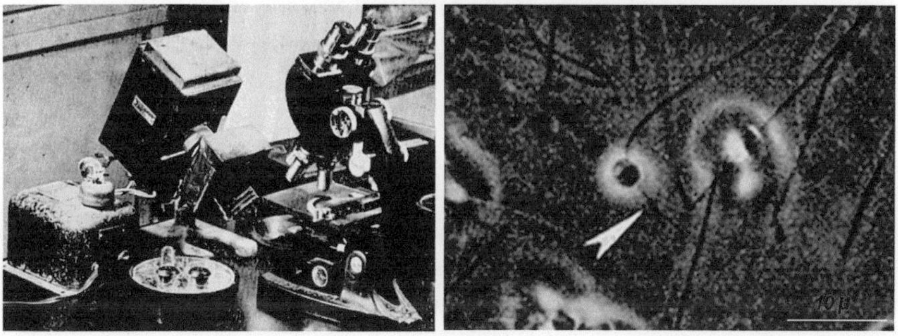

Fig. 2.7a *Left*: The first phase contrast microscope produced in the United States (by Bausch and Lomb Optical Co.). At Ken Cooper's urging, Prof. E. G. Conklin of Princeton University arranged for the purchase of this equipment by a grant from the American Philosophical Society. Jean Dan brought the microscope home to Misaki in 1948 as a present to her husband. But Katy wanted Jean to return to science so he told her to use the scope herself. *Right*: Phase contrast image of star fish sperm with acrosomal process. The white arrow points to the process induced by egg seawater [Jean Clark Dan et al. 1954. *Biol. Bull.107*: 203–218 (Fig. 4 cropped)] (Jean Clark Dan et al, Biol. Bull. 107. 1954)

Zernike was able to design a microscope that converts the invisible image of small cell parts into clearly visible images. Now minute, clear, colorless parts were directly visible in living cells, and their movements, etc., could be followed, owing only to their slightly greater or lower refractive indexes. For this contribution, Zernicke received the Nobel Prize in 1953.

A few years later, after arriving in Princeton, I learned that the unit that Jean had brought back for Katy was, in fact, the first phase contrast microscope that Bausch and Lomb in Rochester, New York, had made commercially. It was donated to the Dans by the American Philosophical Society at the recommendation of E. G. Conklin, as suggested to him by Kenneth W. Cooper, both at Princeton University. It was with Kenneth Cooper that I did my PhD work at Princeton. But let me digress here and convey the two different stories of how Jean and Kenneth had actually met in Woods Hole.

Jean's version is that one evening while she was relaxing at the Captain Kidd Bar, a common drinking place for scientists and students in Woods Hole, she ran into this handsome tall fellow named Kenneth Cooper. She talked with him about Katy's research at Misaki, the crazy young fellow (me) in Katy's lab, and about microscopes. The upshot was a recommendation for me to apply to Princeton for graduate work, and for Kenneth to talk to Conklin about a grant to secure a phase contrast microscope.

The other version, as told me by Suzie Steinbach, one of Jean's closest friends and wife of Burr Steinbach (another fellow student of the Dan's with Heilbrunn), was that she, Suzie, had actually introduced Jean and Kenneth to each other; not that they had met by accident at the Captain Kidd. Perhaps both versions are partly true, but in any event, Jean did bring back the phase contrast microscope, and I applied to Princeton.

Tales of Woods Hole

At Misaki, in 1948, we heard many tales of Woods Hole from Katy as well as from Jean, who had just returned in high spirits from her homeland. Those stories were not just about the encounter that Jean had with Kenneth, but about Katy and Jean's student days in the mid-30s at the MBL, and about their many friends and acquaintances in Heilbrunn's lab at Penn and in Woods Hole. They included Dan Mazia, Don and Helen Costello, Suzie and Burr Steinbach, and Arthur and Laura Colwin. Also at the MBL there was Doc McInnis, who ran the collecting boat, and Jim Graham from Penn, a superb glass blower who was full of juicy, and at times spicy, gossip about the scientists who worked at the Lab. Another long-time US resident was Dr. Tashiro, who lived by the Eel Pond and befriended young Katy.

We also heard about Jean and her friend's canoe trips to Devil's Foot and beyond, the beach parties and about the picturesque location of MBL on the southern coast of Cape Cod.

News of Katy's triumphant 'Millikan's oil drop' experiment using Kunitz's electrophoresis apparatus (made for him by Jim Graham) with which he determined the surface charges of several species of marine eggs piqued our curiosity and made us further excited about MBL.

Jean also told of her experiments with a plant cell, Nitella, in which she measured the deposition of calcium oxalate into the cell sap. To their mentor, Heilbrunn, everything going on in the cell was governed by calcium ions, a notion laughed at by his colleagues, but today turning out to be so prophetic.

For those of us hungry for more science, the MBL at Woods Hole was the true target of our dream.

Jean's Return to Science: The Sperm Acrosome

With the newly arrived phase contrast microscope, Yoshiyuki Endo started exploring the changes at the egg cell surface at fertilization. He was able to show how, upon sperm entry, the fertilization envelope became lifted from the surface of the egg surface by secretion of a gel from the cortical granules into the 'peri-vitellin space.'

Jean, on the other hand, was busy helping out at the Co-op in Nagai, which had grown into a thriving business by then, but that was not to continue for long. One day, on their bike ride to Aburatsubo, Katy challenged Jean to return to her life at the lab full time and start concentrating on biology, possibly using the phase contrast microscope herself. Katy had confronted Jean with a very tough choice. To give up her civic activities and lessons in democracy to her Japanese friends was no easy matter, yet Katy insisted that her science could not progress without total commitment on her part.

Jean opted to listen to Katy. Thus was born her epoch-making contribution. She discovered the 'acrosomal reaction' in 1953 which finally showed how sperm and egg become united as the sperm approached the egg. Before Jean's discovery, the initial interaction between sperm and egg was a mystery. Many felt there was something like an immunological reaction that united the sperm and egg. In the literature, there were also descriptions of a tenuous thread that was said to be occasionally seen between the egg surface and sperm. Some who claimed such observations speculated that the egg threw out a thin thread to lasso the sperm.

In 1952, using the electron microscope in collaboration with Hiroshi Akahori, Jean compared sea urchin and starfish sperm that had, and had not, been exposed to 'egg seawater' for various brief intervals of time. What Jean established was that the sperm reacted to egg seawater and produced a tenuous thread from the acrosomal vesicle located at the tip of the sperm head (Figs. 2.7a and 2.7b). Although extremely thin, this thread, the 'acrosomal process,' was also visible under the phase contrast microscope. By observing live sperm and egg of sea urchin using the phase contrast microscope, Jean found that the sperm reacted to the jelly surrounding the egg (the triggering agent in 'egg seawater') and generated from its tip the extremely thin

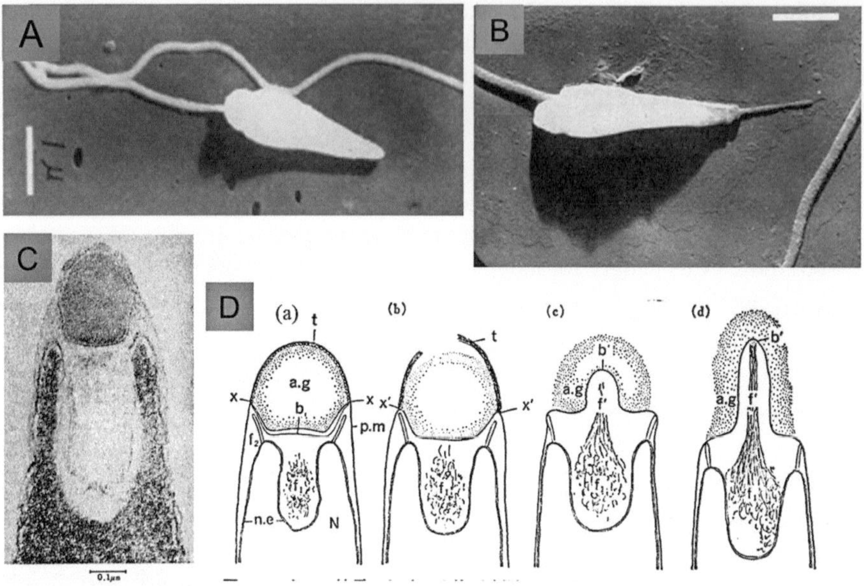

Fig. 2.7b EM of sea urchin sperm. (**a**, **c**) before and (**b**) after acrosomal reaction induced by exposure of sperm to egg seawater. (**a**) and (**b**) are shadow-cast whole mounts, and (**c**) a thin section. (**d**) shows interpretive diagram of early phases of acrosomal reaction (Courtesy H Akahori, Jean Dan. 1952)

'acrosomal process.' The rapidly growing acrosomal process penetrated the jelly to reach the actual surface of the egg cell. Upon fusing with the egg, the sperm was drawn in to the egg via the acrosomal process, often with the intact sperm tail still seen vibrating inside the egg under the phase contrast microscope. Jean, together with her students at Ochanomizu Women's College and Laura and Arthur Colwin (who visited the Dans at Misaki), confirmed Jean's initial observations using live sperm and egg from many different species of marine organisms.

Off to America

Departure

Late at night in August, 1948, we were called at our Magome home that the freighter I was to take tomorrow from Yokohama to Portland, Oregon, would be leaving at the crack of dawn instead of at 10 AM as scheduled. It turned out that the captain not only wanted to ride ahead of an impending storm, but also to sneak out and reach Portland, ahead of his competitors. In those days, the merchant marines were fiercely competitive; the first to arrive at the port capturing what trade was available.

No matter, I was all set to go. I had the invitation letter to attend the Graduate School at Princeton, I had already sold my book collection that included 12 volumes of '*Experimental Physics*' to supplement the loan from the Chitticks, and the new three-piece suit that my mother had ordered tailor made for me was hanging in the closet. At the American Embassy, my visa had cleared just in time, since I was not the same Shinya Inoué who had left the United States for Japan at the start of the War. [It was a complete surprise to me that there was another Shinya Inoué who had been in the States and had left for Japan at the start of the War.] In any event, I was completely ready including with my full-sized chest X-ray picture, required in those days to prove that we were not bringing tuberculosis into the United States.

On board the ship, there were half-a-dozen Japanese students heading for different parts of the States, and one lone student from the Philippines who told me "I'm afraid of you." I could not quite understand him, since although I had heard of the inexcusable atrocities committed by the Japanese soldiers in China and the Dutch Indies, and even to the Japanese themselves in Okinawa (see Cook and Cook 1992), I was not yet aware of the treatment of captured prisoners in the Philippines and elsewhere.

We did beat the storm and got to Portland, Oregon, in 10 days. Then how I got to New York from there, I just don't remember. What I do remember was being appalled at this heavy-set fellow who tried to sell me a diamond ring outside of New York's Grand Central Station; he made a huge scratch on a store window to prove that it was genuine diamond! Ah, what you come across in big cities.

On the Penn Central train from New York to Princeton, I ran into this well-dressed lady sitting across from me, who said she was editor of *Dance Magazine*. We got to chatting about this and that until the conductor came and said, "Why didn't you get off at Trenton to change trains as I told you to?" to which the lady answered indignantly, "But he doesn't need to get off yet to go to Princeton." Ah, such confidence! So I rode all the way to Philadelphia and had to back track to Princeton Junction.

Princeton

Arriving at Princeton, I was assigned a room with a bed under the dormer on the third floor of the Graduate College, the resident hall for graduate students. One of the first people I ran into was a janitor with broken English who told me to be sure to tuck the shower curtain inside the stall. It is strange how this poor Greek immigrant, who was trying to help me from flooding the room, somehow scared me by his advice. I must still have been afraid of any curt authoritative instructions.

A couple of days later, when I returned to my room, a mother cat was trying to give birth to kittens right on my bed! I quickly found an empty cardboard box, placed the laboring cat inside, and dashed down to the basement of the Graduate College just in time for her to deliver her babies there.

In front of the Graduate College, Larch and other trees were scattered on a sloping green lawn, which stretched for perhaps a half mile, as though this were a golf course. To the right was a tree-lined road where Albert Einstein lived, and a little further down, the Institute for Advanced Studies, where Einstein and Hideki Yukawa of the Mu-meson fame worked. The classic Gothic architecture of the Graduate College resembled universities in England, so I used to joke by telling my British visitors from Oxford and Cambridge that they had copied Princeton's buildings.

Dinner at the Graduate College was a formal affair, with Deans Taylor and Thorpe sitting at the head table. Until the year before we arrived, everyone was to appear for dinner with black gowns. But apparently the custom got discontinued when too many students were rushing in after their tennis games, clad with nothing but their gowns over their naked, sweating bodies.

Speaking of characters at the Graduate College, there was a math grad student who thoroughly impressed me. The occasion had to do with his response to a match trick. The trick, taught to me by one of my sisters, was very simple. From a book of matches, take out two sticks and place a prominent ink dot in the middle of one of the two surfaces of each stick. Hold the two sticks suspended lengthwise between the left thumb and index finger, and have the audience verify that the ink dots are clearly on the top surface and not on the other. Grab the middle of the two sticks with the thumb of the right hand on top and index finger on the bottom. Now turn the wrist of the right hand so that the thumb is now below. Return the match sticks, again lengthwise, to the left hand and remove the right fingers. The ink dots which should now be on the bottom surface are still on top!

Observing the actual event, the trick looks totally magical and has baffled countless individuals, but not this math student. He simply stated "You flipped the sticks over as you were twisting your wrist." To my question, "Did you see me do that?" his answer was, "No, but that's the only thing that you could have done." Such simple, straight-forward logic, rather than trusting one's eyes and being fooled like the rest of us had been! This math graduate student turned out to be John F. Nash who won the Nobel Prize in Economics in 1994 for his work on 'Game Theory,' the basis of his PhD thesis at Princeton in 1950.

Another mathematician John Tukey, as a bachelor, used to eat with us at the Graduate College. Befitting his large build, his consumption must have exceeded ours by several fold. This same John Tukey designed the highly accurate bomb sight calculator (undoubtedly analog and not digital in those days) that yielded incredible pinpoint accuracy to the bombs dropped by the B-29s from way up high. At Tokyo University, which was planned, but not used, as occupation headquarters, our science building was unscathed while the stores across the street, only 15 or so feet wide, were all in ashes.

Speaking of dining at the Graduate College, quite often we used to be served fried scallops for dinner. Joe Hoffman, who entered the Biology Department soon after us, posited that Howard Johnson, the caterers for the dining room, would not use expensive real scallops, but that these were skate muscles punched out to look like scallops. My palate suggested otherwise, so I wagered him five dollars that these were real scallops, and that I could prove so by looking with the polarizing microscope. Well, when I took a piece of the 'scallop' to the lab and looked, sure enough the striations did not run regularly across the muscle fibers, but were criss-crossed. In fact I didn't know what scallop muscle, nor skate muscle, should look like. But I took a gamble and told Joe that the striation pattern was that of scallop and not skate.

That won me the five dollars; but only for some 30 or so years. Joe, who by now was President for the Society of General Physiologists, had asked me to give a talk at one of their national meetings. And he relayed what we had experienced as graduate students at Princeton. So as guest speaker, what could I do but to confess that at the time of our bet, I had no idea what scallop muscle should look like. The five dollars that I returned to Joe from the podium generated a hearty laugh form the audience. To this day, Joe still has a hard time believing that a serious student like me would pull such a prank on him.

One evening a week, the College invited distinguished speakers to discourse on a broad range of topics. That is where I flustered Robert Oppenheimer by asking him why the A-bombs were not dropped on some uninhabited island rather than on cities crowded with civilians.

Biology at Princeton

From the Graduate College, Guyot Hall (which housed the Biology and Geology Departments) was a 15-min walk between the undergraduate dorms to the other end of the campus. A few days after arriving at Princeton, I found my way to Guyot Hall and entered Kenneth Cooper's office in the Biology Department, whereupon he exclaimed, "Some goddamn fool twisted my ankle and sprained it last night." What was I to do but to confess, "That must have been me!" Kenneth was referring to the party that had turned rather wild at the senior graduate students' apartment the previous evening. Judy Price, one of our hostesses, and I had decided that it would be fun to crawl under the coffee table two-by-two and started the procession. Several people followed in pairs, but not Kenneth Cooper who was reclining on the sofa and refused to join in the merriment.

Several of us decided that that wouldn't do, so we lifted Kenneth's heavy body off the couch and literally pushed him under the coffee table. I was grabbing on to one of his feet, but in the struggle, his loafer came off and apparently his ankle got twisted. Well, some introduction to my mentoring Professor, but that confession broke the ice, and we became exceptionally good friends – for life (Fig. 2.8). For the first few months after arriving at Princeton, Kenneth arranged for me to go up to New York City once every 2 weeks or so. The doctor, located on Second Avenue (I think it was around 60th Street), helped me regain my full strength by correcting my nutritional deficiency.

Next to Kenneth's office-lab on the second floor in Guyot Hall, where I was to do my thesis research, was Dr. E. G. Conklin's office. Sharing his office was Dr. Ethyl Brown Harvey with her microscope set up on a lab bench placed against the ever-

Fig. 2.8 Kenneth and Ruth Cooper with their adopted children, Geoff and Terry (Courtesy K Cooper)

locked connecting door to Kenneth's lab. As first-year graduate students, we were not allowed to start right into our thesis research but had to take several courses as well as act as teaching assistants for undergraduate courses.

Among the memorable courses at Princeton was Dr. Elmer Butler's gross anatomy course, where, starting at some terrible hour around 7 in the morning, we dissected and sketched a pickled dog fish (a small shark with pliable cartilage instead of hard bone) and then a cat. Surprisingly, I did quite well in this course, but not in Kenneth's advanced genetics course. As mentioned earlier, he thought I should go and take the undergraduate, introductory genetics course to learn the fundamentals. In contrast, Dave Stadler, whose father was a well-known corn geneticist in Missouri, and my other classmate and eventual roommate, Woody Hastings, did quite well in Kenneth's course. I can't remember how our fourth classmate, Charlie Roos, did in this class, but perhaps by then he had already transferred to be a physics major. Of my three classmates, Woody and Dave served in the fields in Europe during WW-II and were supported by GI bills of right.

Different professors held their individual style of classes. Dr. Pringle had us all gather at his home in the evening where he would show us slide after slide of grossly deformed individuals and lecture to us on endocrinology. In those days, very little was known about mechanisms of endocrine action, especially at the cellular level, so much had to do with descriptive pathology and remedial drugs. I'm afraid I didn't much care for those subjects, even with the treat of pretzels and beer at his home.

In contrast, there was Dr. E. Newton Harvey. During his honeymoon in Japan in the 1920s, together with his young bride, Dr. Ethyl Brown Harvey, they discovered how light was emitted from *Cypridina*, a small shrimp-like crustacean, which they collected by suspending sardine heads in the sea. They showed that a substrate, luciferin, was catalyzed by an enzyme, which they called luciferase, and gave off light in the presence of oxygen, thus establishing the first known mechanism for bioluminescence.

The heavyset, jovial Dr. Harvey had a big room full of strange looking instruments in his basement lab of Guyot Hall. In that room, a blackboard covered the whole front wall, and in front of the blackboard sat a 12-foot-long demonstration bench. He would stroll back and forth between the blackboard and the bench and talk about interesting studies in cell physiology, including the work that he and Mrs. Harvey had done with the centrifuge microscope. We four students sat on the other side of the big demonstration bench and took notes. One day, however, I was so tired that apparently I fell asleep with my head right on the demonstration bench. When I awoke, there was no Dr. Harvey or any of my classmates; at the end of class they had tiptoed out of the lab and left me to find myself fast asleep, and all alone!

Aurin Chase taught us biochemistry, but not as a dull list of chemicals and reactions that poor medical students have to memorize today. Instead, he showed us the principles of chromatography, for example, by demonstrating how a drop of black ink (or chlorophyll solution) placed on filter paper spreads out into rings which reflect the different diffusion coefficients of the component chemicals ['black' ink was actually a mixture of red and green ink!] Furthermore, he showed us how to obtain quantitative spectroscopic data by an ingenious use of photography. First you expose stripes of the unknown's spectrum on to a photographic plate, alternating with stripes of the standard spectrum. Then while exposing, you modify the exposures of both sets of stripes together by using a spinning sector wheel which reduces the transmitted light in exponential steps. After developing the plate, you select the particular sectored image of the unknown spectrum which has the same brightness as the sectored image of the standard spectrum. The accumulated choices made in this manner gives you the sought-after values of the unknown spectrum relative to those of the standard. As one who was averse to black boxes, I appreciated such delightful experiments that clearly demonstrated their principles.

I believe it was in the second semester of our first year at Princeton, that we had an experimental botany class with two young instructors, John Bonner and Bill Jacobs. John had discovered the rhythmic migration of cellular slime mold amoebae, which was triggered by their periodic release of cyclic AMP. What he found was that other amoebae responded by migrating towards the released c-AMP while further adding to the concentration by secreting more c-AMP themselves. Thus a whole stream of amoebae collected, as though an inverse vortex, towards the original source. We were fascinated by his time-lapsed movies of the rhythmically migrating amoebae and their role in formation of the fruiting body. John Bonner, who wrote many books on animal behavior, and who married Jim Graham's daughter, Ruth, was another favorite individual of mine, leading to another life-long friendship.

For the lab exercise in this course, we had fun doing whatever project we wished. I recall Dave Stadler showing mutants of a mold and how their 'ascospores' lent them to rapid genetic analysis. At the end of the class, Woody Hastings pushed in a cartful of Petri dishes and reported that, contrary to his wishes, the samples he had collected from the woods had still not shown any bioluminescence. Woody was, however, eventually to become one of the world's foremost authorities on bioluminescence and its circadian rhythms.

As for my exercise in the course, I decided to trace the direction of cytoplasmic streaming in *Elodea* leaves since the pattern looked totally non-random, as though each cell was affecting the streaming directions of adjacent cells. However, by measuring the statistics of the streaming directions at the junction of three cells, of adjacent cells, (and of another parameter I can no longer remember), it became clear that what to our eye looked completely non-random and highly patterned was, in fact, the result of total randomness. By copying the outline of cells and assigning clockwise or counterclockwise rotation to every cell using a table of random numbers, the resulting map indeed looked highly patterned, just as did the natural leaves. This is one of the few times that I felt statistics was useful.

The fourth member of our class, Charlie Roos, was a friendly and generous big fellow, whose Texas accent was difficult to understand for close to a month. He was interested in biophysics, perhaps more physics than was accessible to biological problems in those days, and eventually switched his major to physics as noted before. He was the only one of the four of us with the resources to own his own car, and many a frosty morning we enjoyed the ride he gave us to Guyot Hall. One morning we became tired of waiting for Charlie and decided to pull a prank on him. So releasing the brake of his car, Woody, Dave, and I pushed Charlie's car from the Graduate College down to the bottom of the hill, and then walked on past the brilliant yellow *Forsythia* bushes to the Biology Department. After an hour or so Charlie showed up, quite upset. Not finding his car in the parking lot at the Graduate College, he called the campus police and reported his car missing. At the bottom of the hill, he came across his car and started driving. Then on the way to the lab he is stopped by the police and has a hard time explaining that he is in fact the owner of the 'stolen car.'

Despite experiencing the prank, Charlie let me use his car a few months later to take my first driver's license test. That was in New York City where his parents had a penthouse on Fifth Avenue overlooking the 'yaks' in the Central Park zoo. I passed the test and was even given the opportunity to take his car out for a drive with a date. I don't remember who my date was, but we ended up far away in Brooklyn somewhere, not having been able to get off the Bridge and Highway after my missing a turn, perhaps on First Avenue.

Charlie and Anne, whom he married soon after we graduated from Princeton, are still my good friends a half a century later, as are Woody and his wife, Hanna, and Dave and his wife, Ann.

For my first Thanksgiving, Woody Hastings (Fig. 2.9) invited me to his parents' home in Delaware, where his father managed a local newspaper and also a motel.

Fig. 2.9 Woody and Shinya at Princeton (Courtesy SI)

After his mother's delicious, bountiful Thanksgiving dinner, I was to give a hand with the addition to their motel. We must have been installing fiberglass insulation above the ceiling. My foot slipped off the rafter, and I made a huge hole right in the middle of the newly installed gypsum board ceiling! That was the end of my big help at the Hastings.

In spite of this mishap, Woody and I still shared a room at Woods Hole in the cottage rented by Jean's sister Peggy Chittick. During the last year at Princeton, when I was awarded the Procter Fellowship, we again roomed together. Ten years later, Woody was in charge of the MBL Physiology Course in Woods Hole. He invited me to join as instructor.

For my first Christmas, Dave Stadler invited me to his parents' home in Columbia, Missouri. There I met his lovely family, learned to play charades, and had my first ice skating experience. But what stays in my mind more than the skating is the ice in the wide natural pond formed in the depression of an abandoned quarry. Under the deepest purple sky I had ever seen in my life, and which was much darker than the roof tops and trees, I was struck by the air pockets that had formed below the surface of the ice. Each a quarter- to a half-inch long, the air pockets were perfect hexagons, and their edges all pointed in the same directions as far as the eye could see.

Unfortunately, I did not have a camera with me to document this amazing sight, but that whole pond, many hundreds of feet long, must have frozen as a single crystal of ice! I had read of Faraday's Christmas Lectures for Children at the Royal Society in London, where he demonstrated the beautiful hexagonal and snow-flake-shaped melting pattern formed on the ice block which he placed in the lantern projector, thus demonstrating the underlying molecular structure of frozen water. But to see a whole quarry pond filled with a single crystal of ice; that was truly amazing!

One day, we were sledding on a road near the Stadler's home, which the city had closed off to cars to protect the sledders. That was where I ran in to one of the few racial slurs that I encountered in the States about Japs. He wondered out loud: what was a 'Jap' doing in Columbia? Instantly, however, the young high school student who had made the remark was reprimanded by Dave's older brother Hank. My naivety became further apparent, when Hank's very friendly wife asked me what other discrimination problems I had encountered in the States, and took her by complete surprise by asking her what the word 'discrimination' meant. This reminds me of the story in Colin Powell's biography in which, as a dark skinned person, he had not known of discrimination until he tried looking for housing as an officer of the US Army (Powell 1995).

It was also in Columbia, Missouri while visiting the Stadlers, that I first met Dan Mazia, a close friend and class mate of Jean and Katy's at Penn as well as in Woods Hole. At a party at his home one evening, we got to chatting about cell division and Katy's hypothesis, and I surprised him by pulling out from my wallet a photo of the spindle birefringence that Katy and I had taken earlier that year. At the time, Mazia was not quite convinced yet of Katy's proposal that the mitotic spindle was an integral body whose elongation led to cell cleavage mediated through the rays of the two asters located at the poles of the spindle. But a few years later in Woods Hole, Dan and Katy together showed that the spindle was a real physical entity that could be isolated as an intact body from a dividing sea urchin egg.

It was also in Columbia, Missouri, that I was introduced to Alex Faberge and Barbara McClintock, who many years later won the Nobel Prize for her work with maize (corn) on 'transposable genes.' They were both members of Dave's father, Lou Stadler's maize genetics lab. Alex, a very ingenious scientist who had come from the famous Faberge family in Czarist Russia, showed me stained slides of chromosomes that had been undergoing mitosis in pollen tubes. The tubes growing out of the pollen grains were so thin that the chromosomes had to slide past each other to complete mitosis. Alex told me then that the trick for growing the pollen tubes successfully was to sprinkle the pollen grains on an inch square of cellophane floating on sucrose solution.

Many years later, I called Alex on the phone to inquire why I was unable to repeat the trick he told me about in Columbia. I was interested in watching how the spindle fibers managed to pull the chromosomes past each other, inside those very thin, live pollen tubes. But the tubes would only grow so far and burst before the chromosomes would even form. His reply was disappointing; he himself had not been able to repeat the experience starting the year after we had met in 1948. The only explanation he could think of was that something had changed in the way cellophane is made. [In retrospect, perhaps some Boron got into the cellophane manufactured the old-fashioned way, i.e., before 1948, since we now know that that element is essential for intact pollen tube growth.]

As we entered our second year at Princeton and started on our thesis research, the Korean War became imminent, and we were told to finish up in another year and a half, instead of taking the usual 4 years to finish our graduate work and PhD thesis.

As it turned out, we all managed to finish our theses without being drafted. But before returning to my thesis work at Princeton, I would like to recall my early experiences at Woods Hole.

Finally in Woods Hole

June, 1949. With the spring term ending at Princeton, Woody, Dave and I packed into Dave Stadler's family car and headed north on the Jersey Turnpike, through the West River Drive in New York City, and onto Merritt Parkway up to Milford, Connecticut. Our first stop was in Milford at the Chitticks' in whose cottage in Woods Hole we were to room for the summer. Olie Chittick was an accountant at Sikorsky Aircraft in the nearby city of Bridgeport, while Peggy, Jean Dan's sister, was busy raising her three active sons. It was this family that had given me a loan enabling me to come to study in the States. Fortunately, I was able to pay back the whole loan out of my teaching assistantship and scholarships before finishing Princeton.

After a brief stay at the Chittick's, we continued north through New Haven and Providence, in those days before completion of the National highway system, by winding through the City of Providence for over an hour. Then through Fall River, the back roads of West Haven to Bourne, across the Bourne Bridge, and finally on to Cape Cod. So much sand and scrub pines was my first impression of the Cape as we headed down the high way until we finally reached Woods Hole. The cottage that the Chitticks had rented for several summers was on a steep hillside, a 10-min walk north up the hill from the Marine Biological Laboratory (MBL, or just the Lab).

My first encounter with the MBL was at the mess hall located on Water Street facing the Fisheries Labs. In the mess hall, row upon row of long tables were arranged, where students and Nobel Laureates would sit intermingled, becoming acquainted and exchanging questions and views as they ate. In addition to Albert and Betty Tyler, Steve Kuffler, Keffer Hartline, many other notable figures ate there, since very few scientists owned their own cottages then.

At the mess hall, the cooking was carried out by older students hired by Homer Smith, MBL's General Manager, and we were waited on by pretty co-eds supervised by Mrs. Meyers. After my first meal at the mess hall, an effervescent Evelyn Sclufer from the University of Pennsylvania came and surprised me with a big smile and warm hug, saying how much she had heard about me from the Bonners. Evelyn and her husband Mel Spiegel and I are still good friends to this day. This is where I also met Helen Loening from Bryn Mawr College, with whom I took sunset-viewing walks to the Wabash estate at the end of Penzance Point.

Mrs. Meyers was a stern supervisor of the mess hall. She would come around the tables and, somehow not appreciating the faces I drew on the empty soft-boiled egg shells at breakfast, made a point of smashing them. The first summer we came to

Woods Hole, a ‘Remember Pearl Harbor’ poster still hung at the far end of the mess hall. Woody Hastings protested and asked Mrs. Meyers to remove the poster, but she would not. The next summer, however, the poster was gone.

It was still not too many years after the Pacific War, and Mrs. Osterhoudt, who looked after and cared for me in Woods Hole like a surrogate mother, was literally afraid for me, as she had been for herself during the War. She said her mother in the Meiji era was the first Japanese woman ever to legally marry an American, but that she had made a point of dissociating herself from Japan during the War for fear of the FBI. She was horrified and remarked “you’ll be killed” on learning that I had hitch-hiked to Falmouth to go to a movie with Louise Seki, a young Nisei student. To me, who was living in heaven and not much aware of the interned Japanese on the West Coast, not to speak of the Oriental Exclusion act or the anti-misogyny laws, her concerns seemed rather exaggerated.

Another scientist I got to know better in Woods Hole was Tay Hayashi. I had first met Tay in Columbia, Missouri, while he was a graduate student with Dan Mazia. Tay was an energetic young physiologist, who grew up in Atlantic City with Japanese parents. When taunted during the War for being a Nisei, he would counter by: “Do you want to fight?” Despite his short stature, those taunting him would back down hearing his hearty laughter. Many outstanding physiologists were trained by Tay, who became chairman of biology at Columbia University and at the Illinois Institute of Technology. Today an annual Tay Hayashi lectureship has been established in the MBL Physiology course to honor his research and teaching contributions.

At the MBL, some of us were allotted laboratory bench space in one of the two rooms on the second floor of the Lillie Building that Dr. Parpart, chairman of the Biology Department from Princeton, had rented. He also had a narrow photographic dark room where he let me set up my microscope, so that I could study the dim polarized light images with dark-adapted eyes. In that same room, Dr. Parpart had set up a microscope attached to a video camera, newly developed by RCA, whose research lab was also located in Princeton. While Dr. Parpart was all excited, I was not too thrilled with the images of the *Arbacia* egg that he showed on the video monitor, since I felt that I could see all of the same structures in the egg by looking directly through the microscope.

On moonless evenings, 3-inch-long sea worms would gather around the patch of light coming from the lamp that John Valois hung from the boat dock in the Eel Pond behind the Laboratory. The red female *Nereis* attracted the male worms, which energetically wriggled their bodies as they swam circling around the female, until his white tail half, full of sperm exploded and managed to fertilize the female. John was working as lab assistant for the Osterhoudts, who were interested in the jelly from the *Nereis* eggs and who also suggested I take a look at them with my polarizing microscope.

Through the pol scope, I saw that before fertilization, *Nereis* eggs had a cortical layer, about 7 μm thick, which was full of tiny birefringent granules. When fertilized, or otherwise activated, those granules poured out into the surrounding seawater and formed the jelly, a gel sphere surrounding the egg. The thickness of the gel layer reached 300–400 μm, many times the diameter of the egg itself. In other words, the same material, densely packed into granules in the 7-μm-thick cortical layer, had swollen several thousand times upon meeting the seawater. And as might be expected for a gel layer swelling outwards from a spherical egg surface, the sign of birefringence was radially positive closer to the egg surface (showing that the positively birefringent molecules making up the gel were constrained laterally, and so oriented radially since that was the only direction in which the gel could swell). When one pushed on the jelly, the sign of birefringence changed to radially negative, or tangentially positive, (showing that the molecules in the deformed outer region of the gel were now lying at right angle to the inner ones). This study became the subject of my first report at the summer-end MBL General Scientific Meetings in 1949 (Inoue 1949).

Albert Tyler's Suggestion: The Chaetopterus Spindle

That summer, Albert Tyler from Cal Tech, a good friend of the Coopers, suggested I take a look at the effect of colchicine on the spindle in live *Chaetopterus* eggs. *Chaetopterus* was a strange looking, flabby, yellowish tube-worm (Fig. 2.10) that gave off blue light when disturbed in the dark. Because of this catchy display in the dark room, it became an attraction for evening visits by the co-eds. But these visits

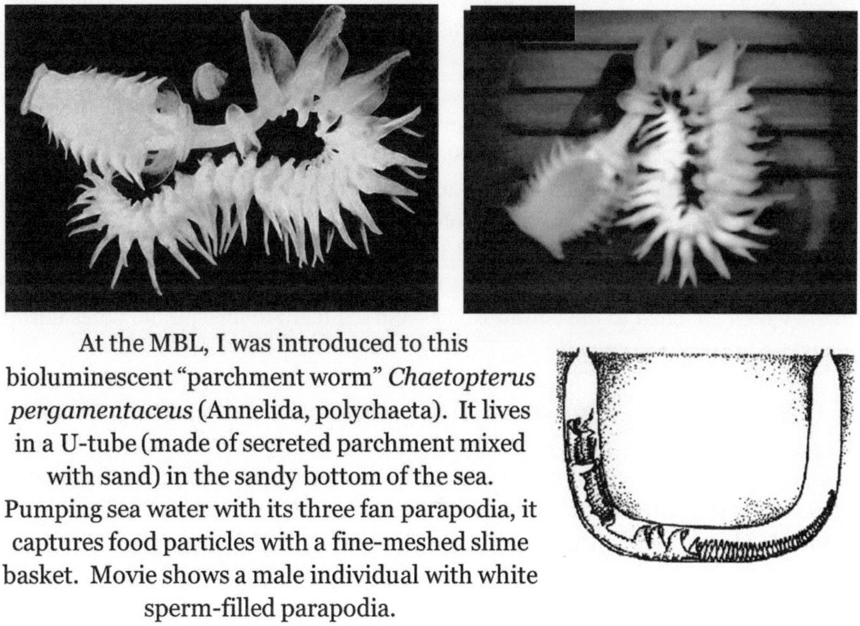

Fig. 2.10 *Chaetopterus pergamentaceus* (Courtesy SI)

to the dark room were soon stopped by Mrs. Osterhoudt, who stationed herself in front of the door "to protect Shinya's important lab work from being disturbed!"

Each *Chaetopterus,* a distant relative of the *Nereis* worm, lives together with a male/female pair of inch-long crabs in a parchment U-tube that the worm secretes in the tidal mud flats. *Chaetopterus* turns out to be an ideal experimental organism for people like me, since all one needs to do to collect the gametes is to snip off a few fat parapodia, the yellow ones on the female worms being full of ripe oocytes, and the white ones on the males being full of sperm. When the oocytes are shed into fresh seawater, their large nucleus (called the germinal vesicle) breaks down, and in just over 10 min, a 25-μm-long first meiotic spindle is formed in its place. Unless the egg is fertilized or otherwise activated, the spindle stays around, locked in metaphase for hours; just ideal for experimental observations of the metaphase spindle.

The *Chaetopterus* eggs, however, turned out to be utterly opaque, being full of yolk and other granules. So I had to obtain clear fragments of the egg that contain the spindle, by stratifying and pinching apart the egg over a sucrose-seawater gradient in an air-turbine centrifuge. After a few minutes' spin in the high-speed centrifuge, the light lipid droplets rose to the top and the heavy yolk granules sank to the bottom of the egg. Between these, the egg was pulled apart like taffy and generated light, medium, and heavy fragments. During fragmentation, the cell membrane would not be torn but sealed itself, so never exposed the cell contents to the outside seawater. Thus, in the appropriate clear fragments, the spindle remained intact. (Exposed to even a small amount of Calcium ions leaked into the cell, the spindle filaments would rapidly depolymerize and lose their birefringence.) In fact, the fragments generated by centrifuging were so intact that upon fertilization their spindle would proceed from metaphase through anaphase, complete meiosis, and would even go ahead to divide the cell.

In these stratified egg fragments free of the light-scattering particles, the weak birefringence of the spindle would show brightly in my polarizing microscope. At Princeton, before coming to Woods Hole, I had improved the scope over the Misaki Shinya-scope so that now I could use an objective lens with higher NA and higher magnification to gain better image resolution.

With this "Shinya-scope 2" (Fig. 2.11; *Collected Works* A-7), I could not only see the presence of the spindle in live *Chaetopterus* eggs, but also the individual fibers which attached to the chromosomes and led them to the spindle poles during anaphase in activated eggs (Fig. 2.12).

The observation with *Chaetopterus* was complemented by views of the forming and dividing birefringent spindles in pollen mother cells of *Lilium longiflorum*, the Easter lily, and other plant cells that I recorded back at Princeton in Kenneth Cooper's lab. Thus, I was able to settle a 50-year controversy and prove that spindle fibers were not artifacts of fixation but were truly present in healthy living cells (*Collected Works* A-10).

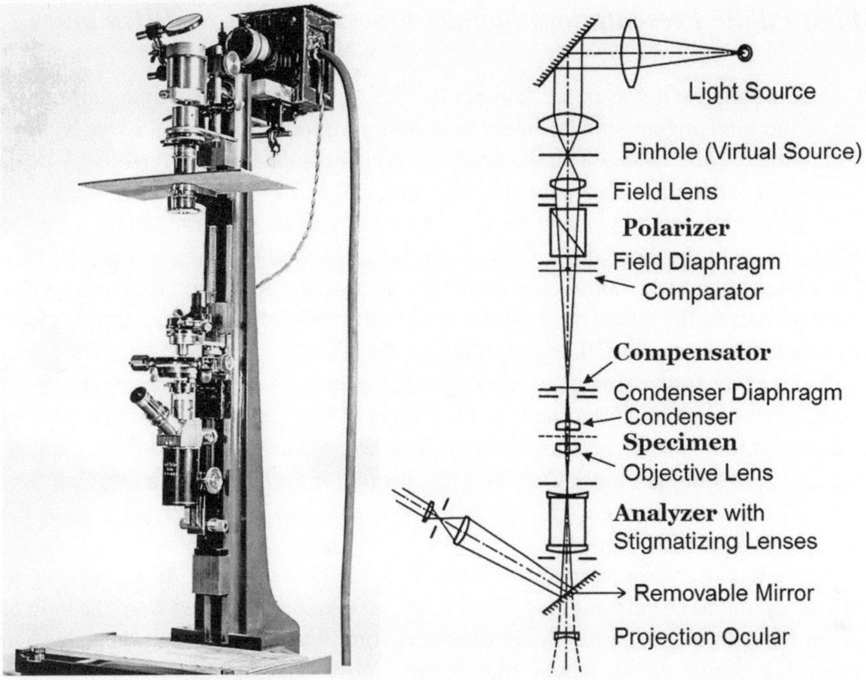

Fig. 2.11 Shinya-scope 2. As a graduate student at Princeton in 1949, I built this improved polarized light microscope (this may have been the first inverted biological microscope). Parts were made available by faculty members of Princeton University (*Collected Works* A-7) (Inoue S, Part 1 Doctoral Thesis. Collected Works of Shinya Inoué, World Scientific Publishing, 2008)

Fig. 2.12 Birefringence of metaphase-arrested meiosis-I spindle in live oocyte of *Chaetopterus pergamentaceus. Left*: Schematic of submicroscopic fibrils deduced from birefringence of live spindle. *Middle*: Elevated birefringence of stretched spindle fibers (in compressed live oocyte observed with Shinya-scope 2.) *Right*: Birefringence retardation vs. length of stretched, or shortening, spindle in live oocyte (*Collected Works* A-10, A-11) (Inoué S, Polarization optical studies of the mitotic spindle. I. The demonstration of spindle fibers in living cells, Chromosoma 5: 487–500, 1953. [A 10]

First Public Presentation: Spindle Fibers, Reality, and Dynamics

One day, I believe it was in the summer of 1951, I was to show the time-lapse movies of the birefringent spindle fibers in dividing *Lilium* cells (Fig. 2.13) as well as slides of the *Chaetopterus* spindle fibers and fibrils disappearing reversibly by treatment with colchicine or with low temperature.

For this presentation, I had to turn to Homer Smith, General Manager of the MBL, for help. In those days, the roof of the Lillie Auditorium was covered by a greenhouse-like glass dome, and there was no way the room could be made dark enough to show the movie or my slides even in the early evening. Homer just smiled and said, "Don't worry, I'll do something about it."

I will never forget Homer's reaction to this request by me, a mere graduate student. What does he do but round up a big sheet of black cloth, climb to the roof of the Auditorium *himself*, and cover up the glass dome. The presentation was successful, and according to Father Cassidy's recollection, I even received a standing ovation. Thank you so much, Homer, but I have more to tell about his role in the development of MBL later on.

The effect of micromolar concentrations of colchicine on the *Chaetopterus* spindle turned out to be very exciting. Colchicine, an alkaloid extracted from the bulb of the autumn crocus, is used widely to collect chromosomes frozen at metaphase in order to examine the karyotype (the chromosomal make up) of cells. It had been

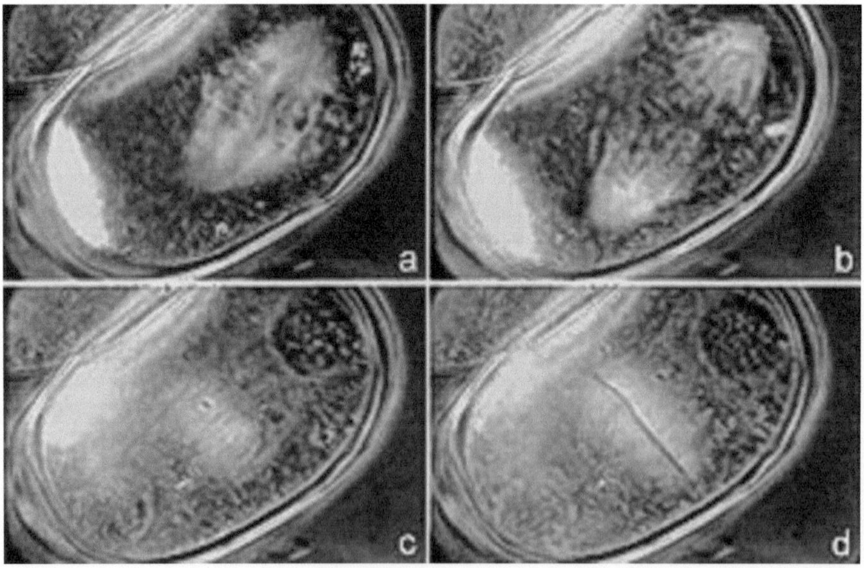

Fig. 2.13 Birefringent spindle fibers pulling chromosomes and birefringent phragmoplast assembling cell plate in pollen mother cell of *Lilium longiflorum* (*Collected Works* A-22) (Inoué S, Organization and function of the mitotic spindle in Allen RD, Kamiya N (eds.), Primitive Motile Systems in Cell Biology, Academic Press, New York, NY, pp. 549–598, 1964. [A 22])

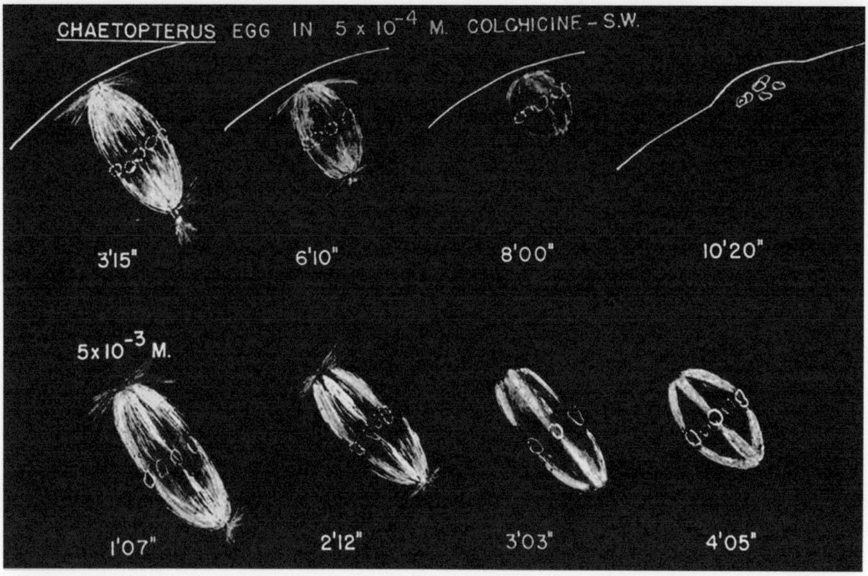

Fig. 2.14 Colchicine-treated spindle. Birefringence changes show depolymerization and force-generating shortening of spindle fiber material (microtubules) in *Chaetopterus* spindle exposed to colchicine (*Collected Works* A-11, A-12, A-58) (Inoué S, The effect of colchicine on the microscopic and submicroscopic structure of the mitotic spindle, Exp Cell Res Suppl 2: 305–318, 1952. [A 11])

speculated that colchicine somehow allowed the cell to enter metaphase but not to proceed beyond metaphase and, thus accumulated the chromosomes stuck in the metaphase stage. (By the way Colchicine, found in the Egyptian pyramids, is also one of the earliest known drugs, which for a couple of millennia has been used to alleviate the symptoms of gout.)

Under the polarizing scope, I was able to show that the birefringence of the *Chaetopterus* spindle faded away in a few minutes in micromolar concentrations of colchicine dissolved in seawater, signaling the breakdown of the spindle fibers and their fibrils (Fig. 2.14). The observation showed that spindle fibers were really needed for chromosome movement, and also explained why the chromosomes could not progress into anaphase but were instead arrested in metaphase.

But what happened as the birefringence of the spindle fibers decayed was truly astonishing. As their birefringence dropped, the spindle fibers *shortened* and pulled the chromosomes and the inner spindle pole to the cell surface, to which the outer spindle pole was apparently anchored. Since the shortening fibers did not get fatter, in fact, if anything they got thinner, I decided that they could not be shortening by the folding of polypeptide chains but were actually *depolymerizing* and falling apart. ***So I concluded that depolymerizing protein filaments could actually perform mechanical work and pull objects!*** When the concentration of colchicine was raised ten times, the birefringence faded away before the chromosomes were displaced. Apparently too fast a depolymerization simply broke up the protein filaments and would not allow them to perform mechanical work.

In the early 1950s, when I made these observations, we knew rather little about the nature of the protein filaments (my 'fibrils') that made up the spindle fibers. In fact, it took another 10 years before microtubules were discovered (Porter 1966), and it was only in 1975 that we could positively show that the optical properties of the microtubules could account for the (form) birefringence of the spindle fibers in living cells (*Collected Works* A-28). Nevertheless, evidence mounted that depolymerizing microtubules, in fact, could induce chromosome movement. Ted Salmon's thesis research in the early 1970s, using hydrostatic pressure as the state variable, confirmed that chromosomes moved faster, the faster the microtubules were depolymerized, and that the movement ceased altogether when the depolymerization was too rapid (Fig. 4.2; Salmon 1976).

But the notion that depolymerizing filaments could perform mechanical work was far too strange for most individuals. They said, "How can you pull yourself up with a rope that you cut?" My response that "we are not dealing with a macroscopic rope, but molecular filaments" would cut no ice for many years. In fact, it took nearly 40 years before Dick McIntosh and others showed that the same phenomenon could be repeated outside of a living cell. They were able to make chromosomes, which were attached to a single anchored microtubule, be pulled towards the anchor as the microtubule was made to depolymerize (i.e., fall apart into their subunit protein molecules tubulin) in the total absence of ATP, a cell energy source (Coue et al. 1991; see also *Collected Works* A-58). So finally, after nearly half a century, my proposal based on observation in living cells that chromosomes could be pulled to the spindle poles by depolymerizing filaments (or more specifically by microtubules that make up the spindle fibers) became generally accepted.

In the meantime, Ian Gibbons at Hawaii showed in 1965 that cilia and flagella beat by the mutual sliding of microtubules that are arranged in neat 9 + 2 rows along their axes. He showed, with an elegant experiment, that the little protein arms attached to the sides of microtubules acted as little oars (using energy gained by hydrolyzing ATP), making adjoining microtubules glide by each other. He called those arm proteins 'dynein' for their ability to generate force. Since dynein and many other 'motor proteins' were soon discovered, also amidst the microtubules in the spindle, most people naturally preferred to attribute the forces for chromosome movement to those motor proteins.

The problem of force generation for chromosome movement is, however, still not settled today, and recent data suggest that both the activity of motor proteins and assembly and depolymerization of microtubules play essential roles in mitosis (*Collected Works* A-58). In a way, it is not surprising that the cell had evolved multiple mechanisms to guard the critical job of bi-partitioning the gene-bearing chromosomes, while also being able to incorporate the new chromosomes that are introduced by mating during meiosis. Albert Szent-Györgyi once quipped in Woods Hole that "cancer research has kept more scientists alive than any other type of research." One could even narrow "cancer research" to "studies on cell division and mitotic mechanisms," since errors in these critical functions do lead to many forms of cancer and developmental anomalies whose general cure still seem quite far out of reach.

MBL Courses and Related Events

Returning to Woods Hole and the MBL of 1949 and the early 1950s, there were many other memorable events. Each morning, before returning to the lab where we had often worked until daybreak, we would attend the Physiology and Embryology Course lectures, which were open to all and attended by students and established scientists alike. In the Lillie Auditorium and in a wooden building called the Old Main (Fig. 2.15), biologists, chemists, physicists, and physicians with different specialties and interests from all around the globe would gather. They contributed to exciting discussions: responses to the apparently 'naïve' but difficult questions from the students, critical arguments from those who had spent years researching related topics, and sometimes even near fist fights.

One particularly contentious discussion took place between Otto Warburg and Ralph Emerson about how many quanta of photons yielded one oxygen molecule as plants engaged in photosynthesis. On the podium in the Lillie auditorium, Warburg shouted at Emerson, "I am only a technician. I do not care about theory, but my measurements show that it takes 4 photons to produce one molecule of oxygen." Emerson in his quiet way had pointed out that the number was closer to 20. So the Physiology Course students (including Wayne Thornburg, who was to become a very close friend of mine) were told to go to the lab and make those measurements using the Warburg respirometer and a green alga (*Elodea* leaves?).

After shaking the Warburg vessels for an hour or so and observing the pressure changes in the vessel due to the evolved oxygen, the students calculated the quantum yield of chlorophyll and came up with a number close to 18. Warburg and his assistant came down from the instructor room in the Physiology wing of Old Main

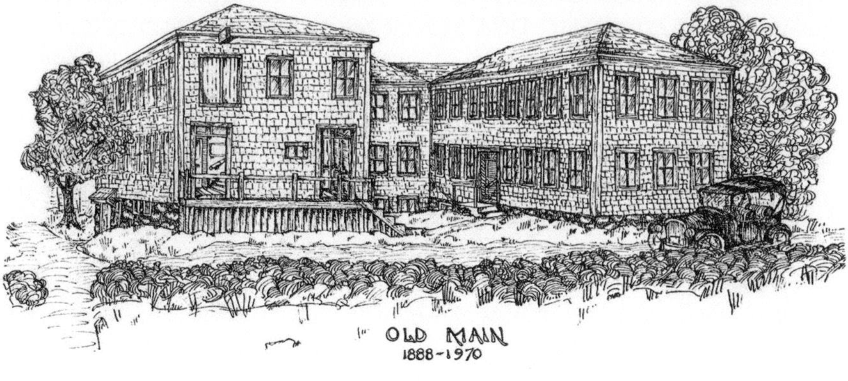

Fig. 2.15 MBL old main building (Courtesy MBL)

and, seeing the results, instructed the students to wait. After a 2-h wait, the assistant finally re-appeared and told the students to re-calculate the data since they had been given the wrong Warburg constant. That constant was nothing more than the volume of the glass Warburg vessel and not subject to much variation, but they tried the re-calculation anyhow and out came the quantum yield of nearly four.

Such an experience was not only eye opening to the Physiology Course students, but to others as well. It turns out that the Osterhoudts had arranged to obtain funds from the Rockefeller Institute to invite Warburg from Germany to Emerson's lab at the University of Illinois in Evanston in order to settle the important question of quantum yield in photosynthesis once and for all. But much to Emerson's horror, and the Osterhoudts' embarrassment, once at Evanston, Warburg asked for the key to Emerson's lab and then locked it to keep Emerson from entering his own room. The performance on the Lillie podium and the episode with the Physiology Course students had followed Warburg's visit to Evanston.

During my second summer at the MBL, a scant 3 years before the Watson Crick model was published, Dan Mazia was lecturing on the importance of DNA as genetic material while also showing DNA's strange physical property. I was much impressed with the gel of DNA that he had isolated from sea urchin sperm. When Dan swirled the beaker containing the gel and then stopped, the gel would unwind itself for several seconds, reversing the direction of the swirling. The DNA molecules making up the gel must be long, long coils that could be stretched for quite a distance and then wind back on themselves like a spring!

Years later, I was told by my good friend Mario Burgos of Mendoza, Argentina, about another experience in the Physiology Course in which he was enrolled as a student that year. Bob Marshak had objected to Dan Mazia's view and argued that DNA could not be an important genetic material since the sea urchin nucleus shows a negative reaction with Feulgen staining, which was well recognized as a definitive test for the presence of DNA.

For his course project, Mario, who was an exchange fellow at Harvard Medical School at the time, was challenged to check out Marshak's claim. So Mario worked day and night for a week until he found that the acid treatment, used to remove the red pigments (in the echinochrome granules) from *Arbacia* eggs prior to the Feulgen staining, had also extracted the DNA from the egg nucleus. By switching to an alternate means for removing the echinochrome pigments, he was able to show the Feulgen-stained nucleus in the sea urchin eggs. That was the proof that Dan Mazia and others needed. [See Fig. 2.16 for Mario who later organized an international video microscopy course in Mendoza.]

Elated but exhausted, Mario took the next morning off to bathe at Stony Beach, where who would appear but Dan Mazia saying, "What are you doing here at this time of day? You're supposed to be in the lab working!"

I believe it was during that year's course discussions in the back room of Old Main that a lanky student, Jim Watson, would be lying on the bench under the window, seemingly asleep. As soon as the lecturer stopped talking, though, Jim would perk up and ask the most penetrating questions. (This sounds similar to the tales about John Tukey, the math professor at Princeton that Kenneth told us about. At the

Fig. 2.16 Mario Burgos (*front row third from left*) at Microscopy Course in Mendoza, Argentina. Ted Salmon (*next to Mario*) and Rudolf Oldenbourg (*back row*) joined Mario as instructors to the course (Courtesy M Burgos)

beginning of each term, one of the headaches for the faculty was to figure out the course schedules so as to maximally accommodate the students who had signed up from different departments. While the rest of the faculty debated and scratched their heads, John Tukey lay supine on the conference table with eyes closed. Finally, he would hop up and a whole workable schedule would unfold!)

More on Woods Hole

But I have strayed way away from Woods Hole again. Despite the long hours that we put in at the lab, Woods Hole was not all work and no play. In addition to the evening strolls to Penzance Point, there was the seawater collecting trips to Nobska Point. With the lead pipes in those days, the running seawater that came into each lab was so contaminated with heavy metal that the sea urchin eggs could not be properly fertilized. So we would take empty five-gallon carboys to Nobska and fill them on the rocks below the fog horn with fresh seawater--a good excuse for inviting along a co-ed.

The fees for using the tennis court at MBL were pretty exorbitant for us poor graduate students so we had to find a way around them. One day Dave, Woody, and I were playing illegally on the Lab courts at Stony Beach when along came the President (Dr. Roberts Rugh?) of the MBL Tennis Club. To his question "Who's a member of the Club?" I blurted out "I am." He asked, "What's your name?" to which I responded "Ann Kaplan." His jaw dropped in surprise, but I think he let us finish our game.

One evening, while I was busy in the dark lab, Dave and Woody had finished their work and decided, at the Captain Kidd with Barbara, Anne, and some other co-eds, that they should go for a skinny moonlight dip at Stony Beach, but on their way there, they left beer cans on top of the cars they passed. The Falmouth Police followed the trail of beer cans and, finding the party at Stony, turned on the spot light. The naked bodies in the water shrank behind the nearest rocks, but the police demanded them to appear. So while the girls still hid behind the rocks, Woody came up to the shore and, as asked, showed his driver's license as ID. The policeman puts his thumb over Woody's last name Hastings and reads out, "John Woodland, aye?" and let it go at that. They were responding probably to complaining phone calls from the neighboring cottages, but I wonder whether the police today would be that good natured?

The stories of our early experience at the MBL and elsewhere in Woods Hole can go on and on (Fig. 2.17). But I should stop here to finish my narrative relating to Homer Smith.

With each new incoming Director to the MBL for the past 20 years or so, I have recounted how things have changed since the days of Homer Smith. Most of the time Homer was managing the whole MBL, assisted alone by Mrs. Florence Butz, his secretary, since the Director of the Lab was in residence only during the summer months. Homer, with Florence's help, would appoint all the student summer help

Fig. 2.17 Friends from the 1949 MBL days gathered again in 2008. *Left to right*: Woody Hastings, Evelyn Spiegel, Shinya Inoué, Andrew Szent-Györgyi, JohnValois (Courtesy MBL)

(i.e., mess hall staff, chamber maids, waitresses, and ground keepers); assign lab and dorm space for all of the investigators, assistants, and students; and arrange for meals and lodging for all of the students and faculty, including their visa and other needs. At the same time, he hired all the regular help and directly oversaw the activities of the buildings and grounds, apparatus, dorms, dining room, and other services functions. He also oversaw the construction of the Whitman and Loeb Buildings that replaced the Old Main and other shingled structures, as well as the Devil's Lane and Memorial Circle cottages and the Swope Center.

While in no way unique to the MBL, the huge growth of the administration over the past few decades has been a major concern to me since it seems to be disproportionate to the services that directly help our research and training efforts. The standard retort has been that federal and state government requirements obligate the increased growth of administrative personnel. Perhaps so, but this is really not appropriate justification; for administration begets administrative boxed-in thinking, an anathema to original pursuits. And the more the administration grows, the more entrenched become the boxes. Despite all these impediments, how can we maintain the outstanding, pioneering quality of science, superseding disciplinary boundaries in research training, original research, and interactions? These have been the hallmark of the MBL, which we must somehow preserve.

Thesis at Princeton

From the fall of 1949 to spring of 1951, I worked on my PhD thesis at Princeton with Kenneth Cooper as my mentor (Fig. 2.8). But before I start talking about the thesis projects, let me first recount some experiences as TA (teaching assistant). Most graduate students at Princeton acted as TAs, earning needed income and at the same time gaining some college teaching experience.

As with most beginning graduate students, my first job as TA was in Biology-1. One gaff that still lingers had to do with my explanation on how to use the light-collecting mirror on the microscope. Soon after the course started, John Bonner who was in charge, exclaimed that a student was complaining about my saying, "Use the *smooth* side of the mirror." Well, the mirror at the bottom of the microscope in those days had two faces that one could choose from, one concave and the other flat. I was trying to tell the students to use the flat side, not realizing that flat and smooth in English referred to very different properties. In Japanese, the same expression "taira" stands for both flat and smooth, e. g., surface of water. I quickly learned to think directly in English.

My most memorable TA job at Princeton was in Kenneth Cooper's Cell Biology Course. In 1950, the term Cell Biology, which is now the name of a ten-thousand-membership international society, was seldom heard. Perhaps Kenneth, who had studied with E. B. Wilson and Franz Schrader at Columbia University, was the one who coined the term. In any event, this lab course is where I myself learned about Koehler illumination and other important aspects of microscopy, as well as much about cytology that Wilson, Schrader, and others had explored.

In the lab, the students and I were exposed to Kenneth's precious collection of original microscope slides made by the pioneering cytologists and histologists. They included E. B. Wilson and Ethyl Brown Harvey's hematoxylin-stained sections of dividing *Lytechinus* (Florida sea urchin) eggs, Franz and Sally Hughes Schrader's slides of mantis and other insect spermatocytes in division, as well as Kenneth's own and Oestergren's slides with which we could follow the unexpected behavior of anaphase chromosomes in grass mite eggs and cells of the plant, *Luzula*. We also got to study David Bodian's beautiful slides of gold fish neurons and their axon hillocks; Bowen's slide showing aggregated mitochondria dividing on the spindle (in scorpion spermatocytes) as though a chromosome; Pollister's slides showing the random segregation of the lipophilic dictyosomes (Golgi vesicles); and L. R. Cleveland's slides showing extraordinary syngamy and divisions in the hind gut symbionts of the wood-eating cockroach, *Cryptocercus*. These are the ones that I still remember vividly.

Even in those historic classical slides, however, there were only a few areas that displayed cells at the right stage, and in an ideal orientation, to illustrate the points of interest, unambiguously. So, in order to locate the cells that illustrated the particular features of interest, Kenneth used a neat trick. He would prepare pieces of index card which exactly matched the size of the microscope slides (usually 1×3 inches in size) and notched them on the upper right-hand corner so that they would always be used in the same orientation. After examining the slide under the microscope and carefully locating the region of interest, Kenneth would superimpose, in exact register, one of the prepared index cards on top of the slide. Then he would mark a circle on the index card where it was illuminated by the condenser since that would show where the region of interest was located. That small pencil-drawn circle, and the drawing and notation also recorded on the index card, allowed him to re-locate the exact spot and points of interest, time after time again. To do so, he would superimpose the relevant card in exact register on the slide, and then adjust their position on the microscope stage until the spot of light from the condenser fell onto the marked circle. All he had to do then was to remove the card and focus the microscope.

Whether this neat trick was a traditional method used by the classical cytologists and histologists of the nineteenth and early twentieth Centuries (who painstakingly studied a number of slides containing numerous microtome sections and laid the foundations for cytology and histology), or whether it was Ken Cooper's original idea, I do not know. In any event, it is a most ingenious and effective approach for re-locating points of interest on a slide, regardless of which microscope is used.

In starting my thesis, the first task was to analyze the limitation of older instruments and build a better polarizing microscope. Kenneth let me take apart his Zeiss microscope, and from Dr. Butler, I acquired a Leitz photocopy stand. For this sturdy, 2-foot-high stand, Russell Mycock in the Biology Shop made adjustable sliders for me—to attach the microscope components at appropriate heights, as well as to support Kenneth's Cine-Special Kodak 16-mm camera for taking time-lapse movies. To generate intense green illumination, Dr. Harvey provided an AH-6 water-cooled high-pressure mercury arc lamp and power supply. I split mica down to a thickness of 5 μm to make my own Brace Koehler compensator; and Bausch and Lomb provided selected strain-free objective lenses. Thus was assembled Shinya-scope 2 (Fig. 2.11). But what of its performance?

After the experience at Misaki, I knew to cross the polars very carefully (to at least a second of an arc with all the lenses removed to achieve the needed extinction) and to make judicious use of the compensator. Then I had to orient the objective lenses to minimize the influence of their strain birefringence since even the best 'strain-free' lenses turned out not to be completely devoid of stress-induced birefringence. But still the field, illuminated with the powerful, concentrated high-pressure mercury arc lamp, could not be fully extinguished.

In searching for the source of this flaw, I checked the appearance of the back aperture of the objective lens by replacing the ocular with a telescope (which is used to adjust the alignment of the annulus in the phase contrast microscope). That telescope also allowed me to photograph the appearance of the objective lens back aperture (Fig. 3.8a).

What I found was that, even with the lenses oriented to minimize the influence of stress birefringence, quite a bit of light was leaking through from the four quadrants between the arms of a dark cross. And the amount that was leaking was worse, the higher the numerical aperture (NA) of the lens. That was very serious since the rays that pass through the high NA region of the lenses are exactly the ones that are needed to provide high resolution of the image. If, in order to improve the extinction, one had to cut out the high NA region of the lens, that meant that one could not achieve the high resolution needed to image cell details if one were to achieve the high extinction needed to gain good sensitivity for detecting weak birefringence.

It turns out that the arms of the dark cross lay exactly parallel to the transmission axes of the crossed polarizer and analyzer. When the analyzer was turned away from the extinction orientation, the cross split into two dark V's which moved out into the areas which were initially bright between the arms of the cross. The further the analyzer was turned, the V's moved out further, in other words, into regions of higher NA. Turning the analyzer in the other direction made the V's move out into the opposite quadrants. Since the dark V's appeared (i.e., the polarized light could be extinguished) where previously there was leaking light, the polarization plane of the leaking 'stray' light was obviously being rotated. The rotation was worse at higher NAs, which is the region needed to achieve high image resolution.

But then, when I started to search the literature, it turned out that this limitation, that precludes the use of polarized light microscopy of weakly birefringent objects with high NA lenses, was already described in 1911 by F. E Wright for petrographic polarizing microscope (Wright 1911). And he had even proposed a partial remedy by suggesting the use of lens elements with lower curvature!

But never mind that Wright was already aware of this limit half a century ago, I continued my analysis by stacking different numbers of strain-free coverslips and/or slides on the stage between the condenser and objective lenses. For the analysis, I used a matching pair of objective lens in place of the conventional condenser lens. Using this arrangement, I found out that the V's would move out further, but retained the same shape as when the analyzer was turned further without the extra glass. In other words, the rotation increased proportionally with the number of air-glass interfaces introduced between the pair of objective lenses. That was to be expected if the 'Law of Malus' held. The law states that the radius of a point at the objective lens back aperture is proportional to the sine of the angle (the NA) that the same ray encounters in the specimen space. Thus, I was able to express the degree of rotation introduced by a particular lens pair as being equivalent to the number of coverslips (twice the number of plane parallel glass surfaces) that were introduced between the lens pair (*Collected Works* A-8).

As we shall see a little later, these studies on the (quantitative aspects of) rotation of polarized light at the lens, coverslip, and slide surfaces eventually led us to find a cure for this serious limit to polarization microscopy. For cell biology, this cure was most important since we needed to achieve both high image resolution AND high sensitivity for detecting weak birefringence at the same time. At Princeton, though, I had not yet learned how to overcome this fundamental limit of polarization microscopy. I had learned what those limits were and went ahead with what was available (e.g., using only a limited NA for the condenser, which would still raise the extinction factor albeit at some sacrifice of resolution) to carry out my biological studies.

In looking for what particular cells would be suitable for studying cell division and mitosis at Princeton, I recalled hearing about Ralph Erickson's remarkable finding that pollen mother cells of the Easter lily divided exactly when the flower bud was 24-mm long. So I visited Ralph at the University of Pennsylvania, where he was kind enough not only to show me exactly which part of the flower bud to measure with a pair of calipers, but also to introduced me to the strange ways that gears were measured (I needed that information in order to design a drive so that I could modify the drive for Kenneth Cooper's Kodak Ciné Special in order to achieve time-lapse recording). Ralph even gave me some lily plants with buds at the right stage from the University of Pennsylvania greenhouse.

Before being able to follow what was going on in the dividing pollen mother cells, there were a couple of problems to overcome. First, I had to devise a medium where the cells would continue to divide for a day or more. (My philosophy, learned from Katy, was that there was no point in studying living cells that were not happy, and that happy living cells alone continued to develop normally.) So I looked for a medium that was isotonic with the cells by checking concentrations of salt or sugar water that almost, but not quite, plasmolyzed the lily pollen mother cells. Then, not knowing what ion ratios the pollen mother cells preferred, I decided to try frog Ringer's solution diluted to 7/8th their normal concentration since I had established that that solution should be isotonic with the cells.

Fortunately, the gamble worked out and the cells seemed happy, but they were so full of birefringent granules that I just couldn't make out the spindle birefringence. So I put the whole flower bud in a simple bench-top centrifuge and spun it for several minutes at around 2000 g (Earth's gravitational field is 1 g). Then I dissected the anthers to let the pollen mother cells flow out into the 7/8th concentration frog Ringer solution.

Now the chromosomes were visible under the polarizing microscope, as were the birefringent spindle fibers that grew encasing the chromosomes. Eventually those fibers gave way to fibers that became attached to the kinetochore (spindle attachment point) on the chromosomes. These birefringent "chromosomal fibers" led the chromosomes to the spindle poles in anaphase. Just as in the colchicine-treated *Chaetopterus* oocytes spindles, the birefringent fibers in the lily spindle slowly shortened as they pulled the chromosomes to the two poles, but without getting fatter.

As anaphase progressed and the chromosomes approached the spindle poles, the initial fibers disappeared, and new bundles of birefringent filaments started to appear in the mid-region of the spindle. Those filaments formed a flattened, foot-ball-shaped body, the phragmoplast, within which many vesicles accumulated, moved to the mid-plane, and fused laterally to form the primary cell wall.

Using the modified Cine-Special, I was able to record this whole sequence for the two succeeding divisions of the pollen mother cell (Fig. 2.13). This time-lapse movie sequence, together with the still images of the *Chaetopterus* spindle, were to *provide the definitive proof of the reality of spindle fibers and demonstrate their dynamic nature*, thus settling a 50-year controversy (*Collected Works* A-10, A-12). I have already touched on this subject when I discussed my early experience with Homer Smith's help at Woods Hole.

At my PhD dissertation exam at Princeton, Jim Hillier, a physicist on the committee from RCA, asked me what happens to the light that is not reflected back by applying anti-reflection coatings on lens surfaces. The question arose since I had reported how anti-reflection coatings reduce the rotation of polarized light at air-glass interfaces and improves the extinction of polarizing microscopes using high NA lenses. (I was aware of this phenomenon since I had convinced Jim Benford at Bausch and Lomb to make what I recall was the first set ever of microscope objective lenses treated with anti-reflection coatings.) Hillier was satisfied with my answer that the quarter-wave-thick anti-reflection film cancels the light reflected from the film-glass interface (since it is now a half wave out of phase with the incoming light), but that the energy is somehow transferred forward so that more light comes through the coated glass surface than the glass surface without the coating.

At the exam, one young instructor at Princeton asked, "Did you design your polarizing microscope in order to prove that spindle fibers were real?" To me this is a shocking question, since I hadn't known what I was going to find. Rather I was stepping into a new world that nobody had ever explored before. Sadly, however, I seem to be a minority in such a view. Today, even most peer reviewers for grant applications seem to prefer sure-fire research, in which the answers are pretty much to be expected, rather than research in which virgin wilderness is explored.

First Encounter with Nymphs

While not directly related to my thesis research, I also had a great deal of fun at Princeton observing some protozoa living in the hind gut of termites. These symbionts of termites digest the cellulose particles that the host ingests, but cannot digest, and convert them into acids that are absorbed through the gut by the host. These strange symbionts are obligatory anaerobes, that is, they need to live in an anaerobic environment and they die in the presence of oxygen, probably reflecting an ancient association with their hosts. L. R. Cleveland at Harvard had called these, and similar (even more anaerobic) symbionts that live in the hind gut of wood-eating cockroaches, nymphidae, or nymphs, that tell us stories. Indeed, these single-celled creatures have many unusual attributes, some of which I plan to relate later in connection with my collaborations with Hope Ritter. (See Part 4; *Collected Works* A-30, A-31.)

Anyhow, the termite hind gut protozoa that attracted my attention at Princeton was *Pyrsonympha*, the pear-shaped nymph. In some textbooks, they were described as having a rigid supporting rod, the axostyle, that runs the length of the cell body and protrudes at its back end as though a tail. Well, under the polarizing microscope, the axostyle was nothing but a static rigid rod. Instead, it was a birefringent whip, which churned inside the cell, beating at the rate of one to a few times a second.

Viewed between crossed polars in the presence of a compensator, it looked as though some material was mysteriously appearing at the base of the axostyle and was constantly flowing down its length. But, in fact, the *Pyrsonympha* axostyle was beating as a saw-tooth wave, with long stretches of straight white regions alternating with short bends that were dark (or vice versa), as the wave flipped down the length of the beating axostyle (Fig. 2.18a). Judging from its positive birefringence, and the fact that the axostyle would sometimes split lengthwise into two or more parts, with each part beating independently as a saw-tooth wave, I concluded that

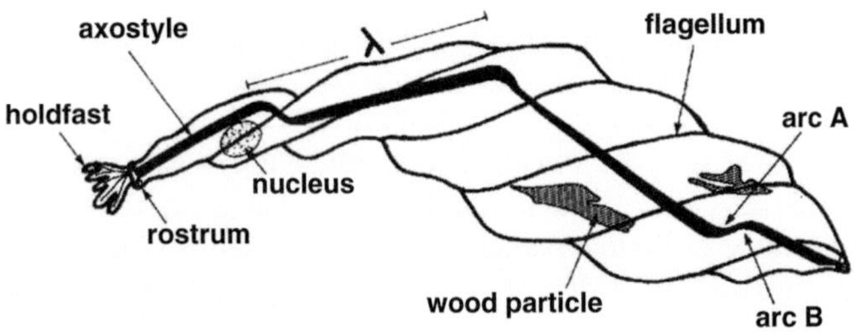

Fig. 2.18a *Top*: *Pyrsonympha vertens*, a wood-digesting, symbiotic protozoan that inhabits the hindgut of termites. The rod-shaped axostyle, originating at the rostrum (at the front end) of the *Pyrsonympha*, is composed of a ribbon-shaped bundle of some 2000 microtubules (Fig. 2.19) and exists entirely within the *Pyrsonympha*'s cell body. When normally active, the axostylar ribbon suddenly forms a sharp bend at the rostrum, then propagates its bend as a twisted, long-wavelength, saw-tooth wave all the way to its tail end, a few times each second. This propagating wave of the axostyle propels the whole organism forward (*Collected Works* A-33) (Gordon GW, Inoué S, Unexpected increase in poleward velocities of mitotic chromosomes after UV irradiation of their kinetochore fibers, J Cell Biol 83: 376a, 1979 [32])

the axostyle was made up of a solid bundle of (protein) filaments, which could bend and twist to form a kink that traveled down the length of the axostyle.

There were at least two intriguing aspects to the *Pyrsonympha* axostyle. The first was that its beating pattern could be made to change dramatically when part of the axostyle was irradiated with blue light. All I had to do was to close down the field diaphragm on the Shinya-scope (Fig. 2.11), so that only a small stretch of the axostyle was illuminated (by the light that came through the still open part of the field diaphragm imaged by the condenser lens), and then remove the green filter. [The ultraviolet and infrared portion of the spectrum from the mercury arc lamp continued to be filtered out by the "heat cut" filters present, so that now the illuminated part of the axostyle was receiving the strong blue spectrum in addition to the green. Filters that removed the green and longer wavelength region of the mercury spectrum did not interfere with the blue light effect.]

What happened was that the propagating saw-tooth wave could no longer pass through the region of the axostyle irradiated with blue light. Instead, the long saw-tooth wave turned into a *stationary* single short sine wave, with the short sine wave staying put at the edge of the irradiated patch until a second saw-tooth wave arrived. Reaching the irradiated patch, that second wave would also turn into a short sine wave and nudge the first sine wave slightly forward. When enough sine waves spanned the irradiated region, a nudge from a new saw-tooth wave would push the original sine wave past the irradiated region where it would suddenly switch back into a propagating saw-tooth wave (Fig. 2.18b; *Collected Works* A-33).

As far as I am aware, the mechanism of this most intriguing effect of blue light on the axostyle (microtubular bundle making up the axostyle) is yet to be unraveled.

This experience not only impressed on me how useful local irradiation of cells by a monochromatic microbeam could be, but on the possible negative effect that blue light could have on living cells. Of course, UV microbeam radiation of cell

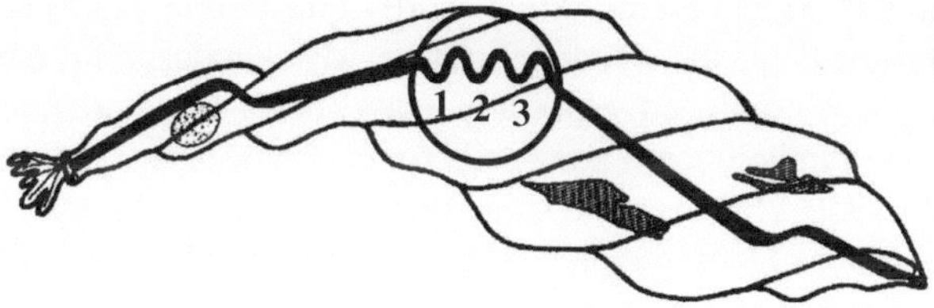

Fig. 2.18b *Bottom*: Sine wave induced in axostyle by blue light. The saw-tooth wave stops when it reaches an area of the cell irradiated with blue light (*circled area*) and transforms into a *stationary,* single sine wave (#1) with a very short wavelength. Then, when the next saw-tooth wave from the rostrum reaches the left-hand edge of the irradiated region and transforms into another short sine wave, it nudges the first sine wave forward (to position #2). Once enough sine waves have accumulated to reach the far end of the blue-light-irradiated patch, the first-formed sine wave instantly converts back to a long-wavelength saw-tooth wave. The saw-tooth wave thus formed becomes self-propagating. From then on, each time a saw-tooth wave from the rostrum reaches the irradiated spot, forming a new sine wave and nudging another sine wave out of the irradiated patch, the tail portion of the axostyle resumes the long saw-tooth wave, which propagates to the end of the axostyle (Courtesy SI)

parts were already used by a number of individuals, but perhaps this axostyle experience was the first time a blue light microbeam was shown to induce a local effect on cell motility.

One day, Kenneth Cooper, in whose lab I was working, introduced me to Fauré-Fremiet, a prominent protozoologist from France who was visiting Princeton. Fauré-Fremiet was not only fascinated with the reaction of the *Pyrsonympha* axostyle to blue light that I showed him, but was kind enough to remember and describe what he saw on this occasion in one of his papers, some 20 years later.

Another behavior of axostyles changed my outlook on electron microscopy completely. That happened about 5 years later when I was at the University of Rochester. In those days, Albert Claude, George Palade, and others were using the electron microscope (EM) at the Rockefeller Institute to describe many aspects of the fine structure of cells, which could now be seen at resolutions at least ten times better than with the highest NA light microscope. In order to preserve those fine structures, they fixed the cells with osmium tetroxide solutions, since it was known that cells after osmium vapor fixation looked remarkably unchanged from life under the light microscope.

So I enlisted Mike Watson, an expert electron microscopist at the University of Rochester, to show me how a fixed axostyle would appear under the EM. What we found in the osmium-fixed thin section of *Pyrsonympha* was just a faint hollow shadow of the axostyle. There was not even a hint of the array of protein filaments that I had deduced should be present from their birefringence observed in the polarizing microscope! Years later, using better fixatives, the axostyle was found to be made up of a parallel bundle of more than a thousand microtubules, tightly packed in a nearly crystalline array (Fig. 2.19).

The next step was to watch the *Pyrsonympha* under polarized light as it was being fixed. As the osmium solution flowed under the coverslip, the axostyle stopped beating, and then over the next several ***seconds,*** the birefringence started fading, until after a minute it was absolutely gone! You can imagine what impact this experience had on both Mike and me. Mike shortly quit biology and started making a new career crafting exquisite wooden furniture. My skepticism for the limitation of electron microscopy continued, especially when people could not find any filamentous structure or spindle fiber in fixed cells for over a decade.

That was in 1950. However, the situation changed with glutaraldehyde fixation introduced in 1963 by David Sabatani of the Rockefeller Institute. Using glutaraldehyde fixation, Keith Porter, who had earlier scoffed at my polarized light observations, gave his famous Ciba Foundation Symposium talk that was published in 1966. He showed that the fibrils that made up the mitotic spindle fibers were microtubules, the same array of protein filaments that made up the axoneme of cilia and flagella (as well as the axostyle in *Pyrsonympha*). Keith even used my polarized light images of spindles in live cells to show that the images he was showing in the EM were not fixation artifacts! To this day, I try to tell people that they should make sure that the birefringence of fixed cells was unchanged from that in the living before completely trusting the EM images.

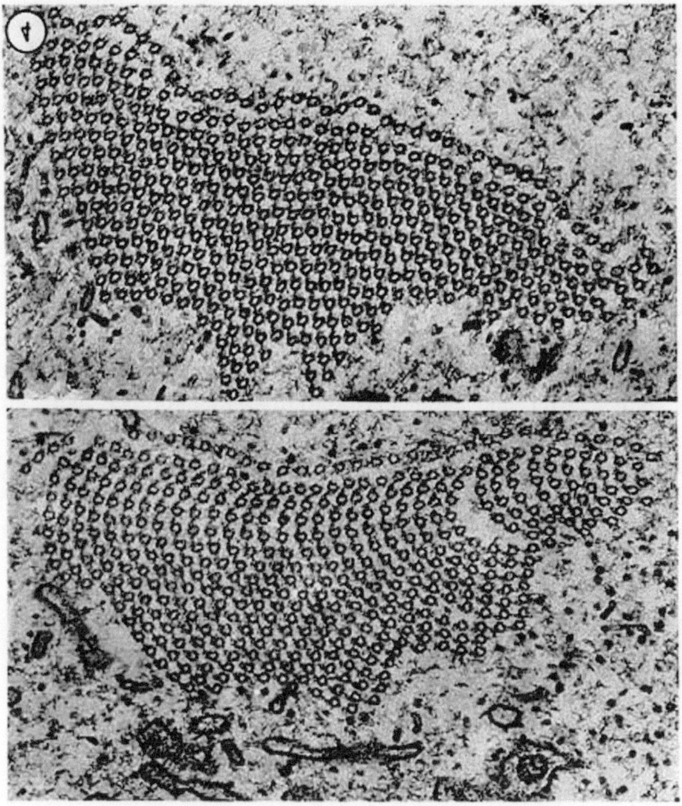

Fig. 2.19 EM cross section of axostyle in *Pyrsonympha*. These microtubules are seen after fixation with glutaraldehyde but not after osmium fixation (*Collected Works* A-33) (Gordon GW, Inoué S, Unexpected increase in poleward velocities of mitotic chromosomes after UV irradiation of their kinetochore fibers, J Cell Biol 83: 376a, 1979. [32])

I've strayed off again somewhat from events that took place at Princeton. But now that I was in my last year of thesis research, and had been awarded the highly prized Procter Fellowship, I roomed in the special suite reserved for the Procter Fellow above the archway at the Graduate College, quite a change from the tiny attic room that I had started out with when I first arrived at Princeton. One day I was happily surprised by a visit from Dr. Bun-Ichi Tamamushi, our chemistry teacher at Musashi. In introducing him to my roommate Woody Hastings, I recalled my poor performance at Musashi, whereupon flustered, poor Dr. Tamamushi blurted out, "Oh no, no. He was a good student!" Well, I guess by then all the math and German courses that I had flunked at Musashi were forgotten.

Now that I was finishing my PhD work, I had to decide what to do next. Katy and my parents had more or less expected me to return to Japan but were willing to let me extend my experience in the States a little longer. My first desire was to go to Harvard in the Society of Fellows program and work with L. R. Cleveland on those alluring nymphidae, but that was not to be. As I learned later, Cleveland was prevented by his Department from taking on any Fellows since he had very few publications to show. Already, in 1951, it was 'publish or perish.' As it turned out, in the next few years, Cleveland gushed out over a dozen historic papers on the symbionts of wood-eating cock roaches.

An alternate and very attractive choice was to go to Cal Tech in Pasadena, where George Beadle had offered me a postdoctoral fellowship. I was all set to go, when Stan Bennett called from his mother's home in Philadelphia. He wanted to talk about the new Department of Anatomy (it was not yet renamed Structural Biology in those days) that he headed at the University of Washington Medical School. With both of us rocking on the swinging bench on the porch of his mother's quaint home, he preached to me about his new Department. He talked about his experience with polarized light at MIT and about his aspirations to introduce more biophysical approaches to cell biology. He had already recruited Wayne Thornburg, who was just finishing his elegant thesis using the microspectrophotometer designed by himself at the new inter-departmental Biophysics Program at the University of Illinois in Evanston. I succumbed to his attractive siren song and decided to join him in Seattle rather than go to Pasadena.

University of Washington

To Seattle

In late summer of 1951, I set off from Princeton to Seattle in the 1927 grey Ford coupe that I had bought for a hundred dollars from the Chitticks. This was the same coupe in which many years ago Jean Dan had driven around the country. The body was not in perfect shape, but the engine still sounded good. As I brought the coupe to Princeton, the seat seemed strangely low, and finding that the legs of the seat had gone through the rusted floor, I added a piece of plywood on the floor to shore up the seat.

My first trip, before taking off across the country, was to Brookhaven on Long Island, New York, where I was to give my first talk at a public symposium. Since I was to head from there directly to Seattle, the trunk of my coupe was full of everything I owned: some clothing, books, slides, and movie film, plus the Shinya-scope borrowed from Princeton for a year. I drove north, smoothly through the New Jersey Turnpike until I started going up the long Pulaski Bridge, when chug, chug, chug, and the engine died on me. I barely managed to coast over the crest of the bridge, then down to get off at the first exit. I landed in a service station in Bayonne, New Jersey, a totally unknown territory to me.

At the service station, immediately three unsavory-looking youngsters saw the padlock on my trunk and started tinkering with it. I had used that large padlock to replace the regular trunk lock which was broken. They continued for a couple of minutes until a man who appeared to be the owner of the station came out and shooed them away. He was a heavy set bearded fellow with only one arm and a patch over one eye, a perfect resemblance to Captain Kidd, the pirate. He asked what the problem was and took a look at the engine for me. His diagnosis: the valve was stuck and I needed a replacement carburetor. He knew a place in Jersey City where we could pick up a replacement and he would drive me there.

When I got into his truck with him, some truck drivers who were resting at the station shouted, "Hey, what's going on there?" However, I had no option but to rely on this fellow. We did pick up a replacement, and back at the station he installed the carburetor for me. I paid and thanked him, feeling how fortunate I was that nothing worse had happened.

As I headed up through New York City and up the Long Island Parkway, it was getting dark. After a while I was looking for my way in a beautiful tree-lined park where I was totally lost. Seeing a police car in the rear view mirror, I flagged him down to ask the way to the Brookhaven National Labs. The officers appeared obliging and suggested I follow them to the office where there was more light and they could explain the road to me. At the time, I had not realized that I had wandered into King's Park, a huge insane asylum built on many acres of parkland, and that the policemen had thought they had caught an inmate trying to escape. But in the lighted office, after verifying my identification and why I was heading to Brookhaven, they ended up escorting me to the highway and gave me detailed directions to my destination.

Even though it was midnight by the time I arrived in Brookhaven, my symposium talk the next day went off smoothly and, in fact, elicited a number of interesting comments and questions. In Brookhaven, I also met a number of people interested in cell division that I had not come across at Woods Hole.

After the successful meeting, the return trip from Brookhaven started out smoothly on the beautiful highway on Long Island. That was until I had to pull off and stop by the highway since I noticed some smoke coming out from under the hood. Having no fire extinguisher, I was at a loss, when a police man in a patrol car, which had been following not too far behind, pulled up and managed to use his extinguisher and put out the engine fire for me. He also found that the fire was caused by a fuel valve that had been inserted backwards when the carburetor was replaced and corrected that error for me. I caught a sigh of relief, but after the officer had left, there was still smoke coming out from behind the glove compartment. Not having any other fluid on me, I urinated into an empty soda can and with it somehow managed to douse the smoldering residue.

Now I was heading for Milford, Connecticut, where the Chitticks would help me get a new battery, new windshield wipers, and a fresh engine tune up. As I was heading into Connecticut from New York, it started to rain. It turned into such a torrent, that almost every other car had stopped on the road side, but I had to keep going. I knew the battery was weak, and I didn't want to get stuck on an unfamiliar highway with no headlights. In any event I did manage to pull into the Chittick's driveway in Milford safely.

I had a couple of restful days with Olie, Peggy, and the Chittick boys at their home overlooking Milford Sound. The car was being tuned up in the meanwhile, and I managed to get in touch with the nurse in Philadelphia who was to ride with me to Seattle. To my surprise, who would answer her apartment phone but Woody Hastings! He said I didn't have to worry about a rendezvous with this nurse in Philadelphia, since he would drive her up to Milford where we could all get together.

So after Woody had driven her up, the nurse and I started off from Milford on our 3000-mile-long trip to Seattle. Connie was a pleasant Nisei girl, but very talkative. And I kept on giving her short answers since I was concerned with whether the engine sounded well enough. After a while I mentioned this concern to her and asked if she could cut down on her chatting. Of course, this was absolutely the worst thing to say, and I ended up having to detour to Yellow Stone Park to sooth her hurt ego. But as fate would have it, the Old Faithful geyser did not live up to her name that day either.

Before we reached Yellow Stone, on the third or fourth day of the trip, way in the countryside in the Midwest out by a farm late at night, the engine temperature started to rise ominously. It was dark, and we were far from any town that would have a service station. Very fortunately I had brought along some spare belts, having read the advice in the *Consumer Reports* magazine. Under the beam of a flashlight and using the tire iron, I managed to replace the frayed and loose fan belt and we were on our way again.

As we hit Eastern Washington State, we ran into a terrible sand storm, blowing straight against us. Before long, the windshield was converted to ground glass, and I had to drive the rest of the way with my head sticking out from the driver's side window. Finally, though, we made it to Seattle, where I was given a room by Connie's parents, the Honda's, who ran a hotel. I was fast asleep in the hotel room, when before daybreak I was awakened by a loud knock on the door. The doorman said "your car is on fire." I told him that couldn't possibly be, after all that I'd already come through, but quickly pulled on my pants and accompanied him outside.

What I found was that a high tension wire had fallen and was hissing, wound around my coupe like a giant snake. Seattle had had a few months of an unusually dry spell, and the rain that night had warped the wooden electric poles to the point of snapping the lines at many places in the city. Of course, it would happen around my car, but it wasn't really on fire.

Fifty miles after arriving in Seattle, my coupe finally gave up. I had to abandon it on the street but was able to recover 50 dollars for salvage value. In those days, the salvage people paid you to tow away the defunct cars rather than the other way around as it is today. So the cross-country trip cost me $50 plus gas and oil, lodging, and meals plus all the adventure. And the car had a life with less than just 2 % of the distance to spare!

The Anatomy Department

For the Anatomy Department at the University of Washington, Stan Bennett had recruited Dick Blandau, an accomplished medical scientist interested in oogenesis, and Ben Everett, who was following the generation of blood cells from their stem cells. Stan had also hired Paul Ralph, a knowledgeable and very kind histologist, who helped Wayne Thornburg and me struggle through teaching the histology lab which was quite unfamiliar to us.

Just before Wayne and I had arrived, Stan had also hired Louise Odor, an analytical chemist, and Lyle Jensen, an X-ray crystallographer. Stan had them help in the gross anatomy teaching, in which, fortunately, he did not involve Wayne or me. Perhaps he felt that chemists were even more unfamiliar with human anatomy than we cell biophysicists, who had at least some experience dissecting vertebrates. Later, Edward Roosen-Runge from Germany, who was an accomplished pianist, in addition to being a pioneer in the developmental biology of zebra fish, and with whose family we became very close, was also to join Stan's Department. Then there was the brilliant MD-PhD student, John Luft. We also had John Roslansky, who would have been my first PhD student had I stayed longer in Seattle.

A few weeks after I settled in, Stan came to my office with a stack of photomicrographs. He wanted to tell me how the 'red flag', a reagent that he had synthesized while at MIT, signaled the location of sulfhydryl (–SH) groups in tissue sections. He particularly wanted to show me how the red-flag staining demonstrated an abundance of -SH groups in the A-band of cross striated muscle. -SH groups are important in cells, since when two of the -SH groups on adjacent polypeptide chains are brought close together and oxidized, they form disulfide bridges. The bridges defined the interaction of protein molecules or of their parts as well as their conformation and function.

When I took a close look at the pictures of Stan's red-flag-stained muscle sections, there was something strange. I could not describe what was wrong so I asked Stan for permission to take a closer look at those sections directly under a microscope. Of course, Stan agreed and had his photographic assistant set up a demonstration for me. They showed me the red-flag-stained slide, alternately illuminated through a narrow-band-pass blue filter and a red filter.

When I pointed out that the focus had shifted at the two different wavelengths, and that the appearance of the cross striations had changed accordingly, Stan was indignant. He said that he was using an apochromatic objective lens on the microscope and that, since apochromats focus all wavelength of light in the same plane, I could not be right. But upon looking through the microscope himself, he grudgingly agreed that there was a focal shift and, by inference, that the images of the muscle tagged with red flag may not show exactly what he had concluded. What a way for a young instructor to challenge his chairman, but I'm afraid this was not the only time that Wayne or I teased Stan.

During spring break, I had my first visit to the University of Washington Friday Harbor Labs. After a couple of hours' drive north to Anacortes, the Puget Sound ferry took us to Friday Harbor after several stops at the other islands in the Sound.

The Lab was a delight. We could collect more than four species of jellyfish right off the dock in front of the lab. Running seawater was piped through Pyrex glass tubing that Art Whitely, Bob Fernald, and Dixie Lee Ray had acquired from a surplus atomic energy plant. The seawater out of the pipe was so pure, and free from excess heavy metal ions, that not only sea urchins, but 100 % of the jelly fish eggs, could be fertilized. As I had learned in Misaki, we collected the jelly fish the night before and kept them overnight in sea-water-filled beakers covered by a dark box. Next day, when we were ready, we would remove the box and expose the jelly fish to room light. Within an hour, the males would spawn and the females would follow shortly thereafter.

One day, the fertilization rate of the jelly fish dropped to less than 20 %, and only 85 % of the tougher sea urchin eggs would fertilize. Suspecting that the buildings and grounds crew had been told by the Lab Director to switch to the old galvanized pipe to flush out one of the seawater pumps, Bob Fernald and I went to look at the pump house. Sure enough, the intake seawater was routed through the old pipes. So that day, I had an oceanography graduate student, who was analyzing trace elements in seawater, collect some of the water out of the tap and compare it with fresh seawater that had not run through the old system. He quickly found that the zinc and copper content of the suspected water was several times higher than in fresh seawater.

Since I had been told that the Friday Harbor Lab Director, who was an oceanographer, did not believe that biologists knew what they were doing by installing the all-Pyrex system, I wrote him to relay the fertilization experiences and results of the seawater analysis. When I returned to Seattle, my department chairman, Stan, told me that such things should never be put in writing. I should have caught him at a cocktail party or some other relaxed occasion and gently hinted at what we had found. Well, perhaps that is the way to succeed in academic politics, but I was not overly impressed by Stan's advice.

In the Medical School at the University of Washington, we had access to an excellent instrument shop (undoubtedly set up at Stan's urging), where we could get professional help in the design and fabrication of precision instruments. Instead of the old Leica photo stand that was the base of my microscope borrowed from Princeton, I was introduced to a Ford spring tester base, a 3-foot-high sturdy cast iron support with a 4-inch-wide dovetail running its full length and supported on an equally sturdy U-shaped cast iron base. The microscope components could now be aligned by sliders, whose height could be adjusted in precise alignment on the rigid, long dovetail. The polarizer and compensator were mounted on the double-spider designed by Stan while he was still at MIT, with the orientation of each indicated via vernier scales to a minute of arc.

Onto the horizontal surface of the spring tester base, I had the shop add two V-grooves running at right angles to each other and a flat. Using this kinematically designed supporting surface, I could simply set down the modified Ciné-Special or 35-mm camera in exact register with the microscope axis.

For time-lapse recording of the faint polarized light images with the Kodak 16-mm Ciné-Special movie camera, I introduced a sectored mirror, tilted at 45° to the microscope axis. This sectored first-surface mirror was made to turn in its own plane (slowly for time-lapsed recording) by a geared-down motor which also drove the film cassette. The sectored mirror alternately sent 100 % of the light beam to the film or to the eye piece so that none of the dim imaging beam would be lost in either path.

Several years later, at Dartmouth, I was introduced to the Arriflex 16-mm movie camera, which had a sectored mirror built in between the lens and film plane, so there was no more need for me to use the film chamber of the Ciné-Special coupled to the home-built sectored mirror system that I had devised.

In any event, this became Shinya-scope 3 that we used in the basement of the Medical Science Building at the University of Washington. We chose the basement location since it was much less prone to vibration than up on the fifth floor of our new building.

If I remember correctly, Wayne Thornburg (Fig. 2.20) was assigned a basement room next to mine, where he set up his 8-foot-long steel I beam, which was the base of his horizontal microspectroscope. This was a very sophisticated instrument that Wayne had designed while a graduate student at the University of Illinois. He even designed the unique reflecting condenser and objective lenses that could focus light of any wavelengths from 200- to 650-nm wavelengths into exactly specified planes (Fig. 2.21).

To obtain the absorption spectrum of the specimen, Wayne measured the weak monochromatic light transmitted through the specimen with a photomultiplier whose quantized output he connected to a pulse counter. This was probably the very first application of pulse counting in microscopy. Furthermore, he devised a chamber for measuring the absorption spectrum of the specimen chilled to liquid nitrogen temperature. Wayne had reasoned that spectral curves obtained at room temperature

Fig. 2.20 Wayne Thornburg (Courtesy SI)

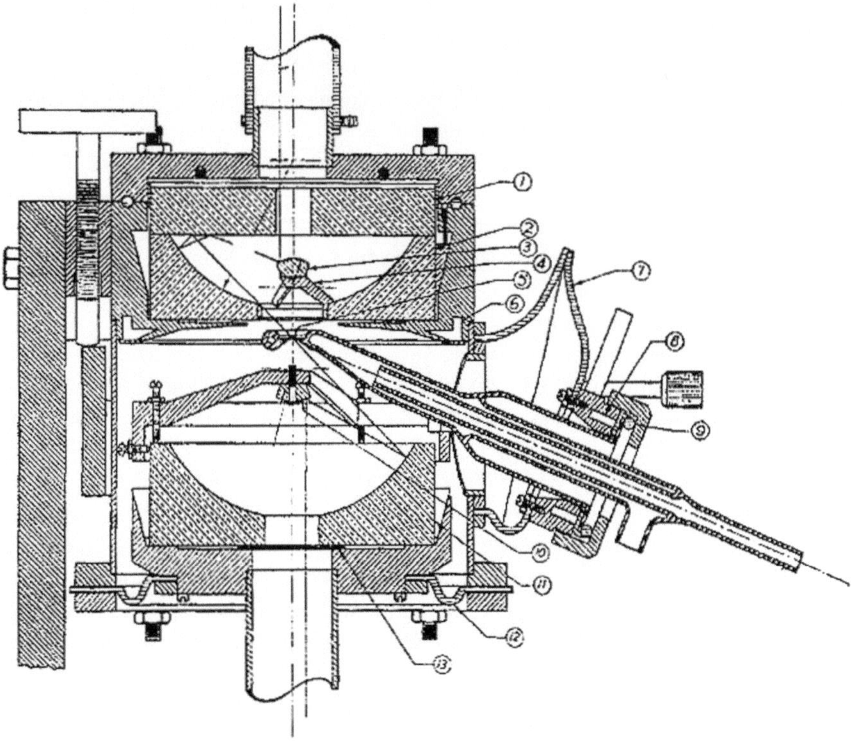

Fig. 2.21 Cross-sectional view of Wayne Thornburg's low-temperature microscope. 1. Correction plate. 2. Concave mirror of microscope objective. 3. Convex mirror of microscope objective. 4. Invar steel tripod. 5. Top of low-temperature stage. 6. Water jacket for thermostating objective supports. 7. Rubber bellows, sealing stage to microscope chamber. 8. Water Jacket for thermostating stage support. 9. Stage clamping screw and rubber seal. 10. Convex mirror of condenser. 11. Concave mirror of condenser. 12. Rubber diaphragm. 13. Condenser aperture stop (Thornburg 1955) (Courtesy W Thornburg)

were broadened by thermal fluctuation of the chemical bonds, and that measurements at a very low temperature where the bonds were much quieter would sharpen the spectrum and yield much better finger prints of the resonance structures. Indeed, as shown in his thesis published in the Journal of the Optical Society of America, the hyper-sharpening of the absorption spectrum of anthracene crystals obtained at liquid nitrogen temperature through Wayne's microscope was spectacular.

MIT Workshop

In 1955, F. O. Schmitt invited me to a course that he organized at MIT. That was an unusual course in that he had youngsters, such as Andrew Szent-Györgyi, Hugh Huxley, Alan Hodge, Betty Geren, and myself, as instructors, while accomplished professors, such as Bill Bloom who had already authored the definitive histology text *Maximov and Bloom*, among the 'students.' Andrew Szent-Györgyi spoke about his analysis of myosin from the muscle A-band, Hugh Huxley about his ingenious new microtome and the ultra-thin sections of striated muscle that he was studying, Alan Hodge about his muscle EM work, and Betty Geren about how the insulating sheath of myelinated nerve was generated by Schwann cells wrapping themselves around and around the nerve axon like a jelly roll. We also learned from Bill Bloom about his work at Chicago on the effects of UV microbeaming chromosome arms, an unforgettable presentation that led to our collaboration nearly 10 years later after I had moved to Dartmouth (see Part 3, Sperm DNA, this volume).

That was the first occasion that I came to know Hugh Huxley who I believe was a postdoctoral fellow in Schmitt's lab. But I had already known most of the other young MIT folks in Woods Hole, since Alan Hodge, Betty Geren, as well as Carolyn Cohen, were in Schmitt's summer lab at the MBL. They were preparing cytoplasm isolated from nerve for analysis, by dissecting out a length of the squid's giant axon and squeezing out the axoplasm like toothpaste, using a glass rod as a miniature rolling pin.

Andrew Szent-Györgyi, or Czuli as we call him, together with his wife, Eve, had been rescued from communist Hungary by his cousin Albert Szent-Györgyi, who was working at the MBL after winning his Nobel Prize for discovery of Vitamin C. Albert, or Prof as he was generally called, managed to sneak Czuli and Eve out of Budapest, by supposedly having them visit Buchtal's lab in Denmark but, in fact, having them come from there to the United States. They arrived in 1949, the first summer I was also in Woods Hole. To our mutual surprise, Czuli said that he had seen my sisters in Budapest, carrying their cello and violin on their way to music lessons.

Another person from Schmitt's lab that I encountered a little earlier in Woods Hole was a graduate student named David McCullogh. He too was studying the mitotic spindle as I was, but using all kinds of fixatives to compare their actions. His PhD thesis was quite a tome, including observations he had made on the birefringence of sea urchin eggs observed after centrifugation and fixation. Unfortunately, he passed away at a young age while he was a postdoctoral fellow with Paul Weiss in Chicago.

Harold Wyckoff was another bright graduate student I met while at Schmitt's course at MIT. He showed me his model of the double helix of DNA, which he had derived from X-ray diffraction patterns obtained from stretched gels of DNA. He also showed me how the X-ray pattern agreed with the diffraction pattern of light waves that had passed through pinholes corresponding to the projected image of his double helix model. Later, I believe, he inaugurated the Biochemistry and Biophysics Program at Yale University.

Speaking of deciphering the double helical molecular structure of DNA, or 'Breaking the Code', the bitter disappointment by Irvin Chargaff (who had figured the important base ratios in DNA) for not having been included in the Nobel Prize is legend. So also is the omission (although apparently unperturbed) of Rosalyn Franklyn, who had carried out the crucial X-ray diffraction work. Few are aware, however, of the unpublished work by Wayne Thornburg, another genius who had investigated molecular models of DNA. Before coming to the University of Washington in 1951, he examined the triple helical model (which was shortly published by the Cal Tech group), but Wayne had already dismissed that model by calculating that it had the wrong density. However, to Wayne, this was just another mental exercise.

Wayne Thornburg and I interacted a great deal in Seattle, and he was the first faculty member that I was to recruit when I became Chairman of Anatomy and Cytology at Dartmouth Medical School in 1959. That is, however, quite a ways off, and it is time to return reminiscing on my personal life in Seattle.

A New Life

Soon after I arrived in Seattle, I was introduced to a number of new friends, many of them Nisei fellows and girls, as was my cross-country travel mate. Among them were Gordon Hirabayashi, Chizu Norton, and Connie Honda, a friendly and cheerful group, many of them Quakers. I also learned about their internship during World War II. While not well known in Japan in the early post-war days, many of these youngsters and families were summarily rounded up and lost much of their assets, because of their Japanese ancestry, and placed in internment camps in Arizona, Wyoming, and elsewhere away from the West Coast. This shameful treatment of loyal American citizens was finally recognized by the US Congress in 1988, after many years of challenge by Hirabayashi and others.

Seattle is where I also came to learn of the 442 Regimental Combat Unit of Nisei's, one of the most decorated units in U. S. military history, who heroically fought with huge losses to save the Texan 1st Battalion of 141 Regiment at the Italian front during World War II. Daniel K. Inouye, who eventually was elected US Senator from Hawaii, lost his right arm in that battle.

One day, Connie, who had labeled me incorrigible (a word I had to add to my vocabulary), introduced me to Sylvia McCandless (Fig. 2.22), who had a 3-year-old daughter, Heather. I soon became very close friends with her. Often, we would talk late into the night on all kinds of topics, not just about our families and experiences, but about world affairs and people and music. During lunch hours we would meet at the herbarium across the bridge from the Med School campus and not far from the University Print Shop where Sylvia was working as a proof reader.

We often went to the Chinese restaurant on University Avenue, run by good friends of Sylvia's, Mamie and Raymond Chin, who were brother and sister of our own age. There seemed to be quite a difference in the relationships on the West Coast among people of different races and background, at least compared to an experience I had in New York. One day, while I was still at Princeton, my father's good friend Mr. Yanagida, head of the Bank of Japan, invited me for lunch at a Japanese restaurant in New York City. We were chatting away and I was describing

Fig. 2.22 Sylvia McCandless (Courtesy SI)

how similar I felt the Japanese and Chinese were in many ways. A Japanese waitress overheard my remark and angrily denounced, "We lost the war because of fools like you!"

A few months after we had met, Sylvia had me take a daylong boat trip up the Columbia River with her father Kenneth, who was Editor of the Pacific Northwest *Farm Cooperative Newspaper*. I recall the huge hydroelectric dam project that he showed me, and his concern for the salmon that needed fish ladders to reach their spawning grounds. We shared interests in photography and soon became good friends. I also had a chance to have Katy meet Sylvia. This was Katy's first post-war trip back to the States, and he was in a hurry to get to Tennessee to collaborate with Gordon Carlson on unequal division in grasshopper neuroblasts, but we had just enough time to taste Sylvia's home-made devilled egg salad. Those were the most delicious devilled eggs I have ever tasted.

Early that summer, in 1952, I drove from Seattle to Woods Hole in my new Plymouth, together with John Roslansky and a couple of other passengers. John and I took turns driving, and in just two and a half days, we reached his aunt's place in Minnesota. I was not aware at that time that John's aunt's family, who had a big car dealership in town, had rushed us to their lake-side cabin, not just because that was a pleasant place to rest up, but John told me years later it was because they had not wanted a Japanese visitor in their home in town.

In a total of five days, we made it all the way to Woods Hole from Seattle. When we arrived behind the Lillie Building, John had us wash my car before doing anything else. I suppose that was natural for someone who had been brought up in a family owning a car dealership, but my strange feeling was confirmed when Evelyn Sclufer (now Spiegel) passing by commented on our activity, which was so untypical for young scientists at Woods Hole.

By mid-summer, Sylvia and I were missing each other so much that I proposed to her over the phone. Without too long a hesitation, she accepted, and we decided to meet at the Logan Airport in Boston where Sylvia would arrive together with Heather. Somehow I got the arrival time wrong, and by the time I pulled into Logan after over an hour drive, both girls had tired of their long wait and had changed into casual clothes. So that was how my bride to be and Heather were dressed when I finally reached Logan. To Eve Szent-Györgyi, who was at a beach party waiting for us, the arrival of the three of us in not the best of moods after the long wait and sweaty drive appeared too comical to maintain a straight face.

In any event, on July 18, 1952, we were safely married on the lawn of the Presbyterian Church in Falmouth by Reverend Long. Katy and his daughter Marina, Woody, Dave and his sister Joanie, Evelyn Sclufer and Ann Kaplan, Tay Hayashi and his student Betsy Clark, Czuli and Eve, Laura and Arthur Colwin, Helen and Don Costello, and Suzie Steinbach were among those attending. Also Olie and Peggy Chittick who took Heather to the beach during the ceremony and returned for the reception. Dr. Parpart from Princeton had a reception for us at their home, and we honeymooned in the windmill which belonged to old Robert Chambers. It was a romantic but primitive abode; we had to stroll to the Chamber's big house even to use the bath room. When I called my mother in Tokyo, her question was, "Are you happy?" Of course, I was up in seventh heaven. After a week, Sylvia had to fly back with Heather to her job in Seattle.

That summer, after visiting Carlson in Tennessee, Katy had come to Woods Hole to work with Dan Mazia. They wanted to show that the mitotic spindle was not just a pattern in the cell but was an actual physical entity that could be isolated as a single body, as demanded by Katy's hypothesis on cell division. In fact, they succeeded by fixing dividing sea urchin eggs with cold 30 % ethanol, then treating the fixed eggs with hydrogen peroxide to vulcanize the protein (oxidize the –SH groups to form -SS- linkages), and then by extracting the still-soluble components and dissolving away the cell membranes with Duponol, an early detergent made by DuPont.

At the end of that eventful summer, Katy suggested we drive out together in my Plymouth to California. I agreed, especially since Sylvia and Heather could come down and meet me in Berkeley, and we could then have a leisurely trip up to Seattle after seeing Katy off. On our cross-country trip to California, Katy would occasionally drive to relieve me. Once, when I awoke from dozing off next to him, he shocked me by saying, "This car can go pretty fast. I got it past 90 miles an hour while you were asleep." He was still full of mischief! After a brief pause at Yosemite National Park, we pulled into San Francisco where Dan Mazia was waiting for us. Young Marina Dan, who with Katy had missed their ship out of Honolulu on their way to the States, wanted Dan Mazia's assurance that that was not going to happen again on their way back to Japan.

After Sylvia and Heather arrived, Dan Mazia took us out for a delicious Japanese dinner before our new family headed up north. Our first stop was at a motel just north of San Francisco. It was in a quiet secluded woodland, situated in front of a natural pond. When we awoke after a good night's rest the next morning, little Heather gave us quite a start by being on the un-fenced deck, observing the minnows swimming in the pond below. Our drive took us through the Sequoia National Park where we marveled at the forest of the millennia-old Red Woods, along the coast of Oregon, then Portland, and finally to Seattle. That leisurely drive up the Pacific Coast felt more like our real honey moon.

Back in Seattle

In Seattle, we started our new life at Union Bay Village, a complex of duplex houses provided by the University of Washington for young faculty families such as ours. A 10-min drive took us to the campus, and a 20-min drive in the other direction took us to Wayne Thornburg's cottage fronting Lake Washington. We spent many weekends watching the ducks as we sailed in Wayne's dingy, and chatting late into the evening at his cottage.

One day that fall, Dan Mazia (Fig. 2.23) called about bringing his isolated spindles to see what they looked like under the polarizing microscope. Of course, I welcomed the idea wholeheartedly. A couple of days later, we looked at his isolates

Fig. 2.23 Katy Dan and Dan Mazia (Courtesy SI)

using my new scope, installed in the basement lab of the Anatomy wing. I first managed to photograph his isolates, which did look quite similar to spindles in living sea urchin eggs, except with the birefringent fibers appearing somewhat coarser. Those pictures appeared in the 1952 PNAS paper, published by Mazia and Dan describing the isolation of '*the mitotic apparatus*' to document the similarity of the isolates and the native spindle in living cells (Mazia and Dan 1952).

Next, we agreed to see how colchicine would affect those isolates. To me the dynamic response of the spindle fibers to colchicine, and to cold, was a true sign of their nativeness. I added a drop of 10 μM solution of colchicine dissolved in seawater to the edge of the coverslip, but a couple of minutes after the solution reached the spindles, the electric power went off. We were literally thrown into the dark until we found our way out of the basement with a flashlight. However, Dan had achieved what he had come for and happily headed back to Berkeley with the photos of the birefringent isolated spindles.

Figuratively, it turns out that the episode had thrown us into the dark for 20 years! It was not until 1972 that a young Assistant Professor Dick Weisenberg at Temple University in Philadelphia, who had been a student with Ed Taylor at Chicago, finally figured out how to isolate ***labile*** microtubules and spindles. Up to that point, spindles isolated by the method devised by Mazia and Dan, and subsequently by others, would neither respond to cold nor to colchicine. Skeptics argued that my observations of spindle fiber behavior in living cells were not showing the direct effects of cold or colchicine (on the molecules making up the spindle) but were the result of some indirect action of those agents on the cell. In a way those arguments were not surprising because the cell is such a complex chemical machinery, but I continued to argue, often in my Physiology Course lectures at Woods Hole, that the isolates made so far had been vulcanized, and that the isolation procedures had depended on partially fixing the labile molecules in place.

What Weisenberg showed was truly epoch breaking (Weisenberg 1972). Realizing that native microtubules (which by then were fairly well established to make up the bulk of the fibrous components of the spindle) were exquisitely sensitive to calcium ions, Weisenberg ground up brain tissue containing microtubules in media that contained EGTA which would sequester calcium ions. If cells were ground up in a solution not containing EGTA, the labile microtubules would completely fall apart since there are a thousand times more calcium ions sequestered in vesicles in living cells than are needed to depolymerize the microtubules. Inside a healthy living cell, the calcium ions are sequestered in vesicles but are released when the cell is damaged or ground up. (Of course, today we know that controlled release of minute amounts of calcium ion from the vesicles play an important role in the control of many cellular activities.) And it turned out that spindles and microtubules isolated in Weisenberg's medium, containing a calcium chelator, EGTA, together with GTP, magnesium ions, and a neutral organic pH buffer, were ***truly*** labile. Spindles and their microtubules isolated in his medium do depolymerize in colchicine or when exposed to cold!

In the meantime, Gary Borisy, who was also a graduate student with Ed Taylor at the University of Chicago, had discovered the abundant presence of colchicine binding protein, i.e., microtubules, in brain tissue where little was supposed to be present (since brain cells were believed not to divide and, hence, not to contain mitotic spindles, believed then to be the main source of microtubules). Shortly thereafter, Borisy and his co-worker Joana Olmsted confirmed that microtubules depolymerized into their subunit protein, tubulin, when exposed to cold in Weisenberg's solution, and that the tubulin molecules would even re-assemble in the test tube to make new microtubules when the solution was warmed up again (Borisy and Olmsted 1972). From then on, chilling cell extracts containing labile microtubules in Weisenberg's solution, centrifuging the chilled solution to remove cell debris, and re-warming the solution to reform microtubules, became the standard method for obtaining pure microtubules and tubulin.

So, yes, the earlier isolates were indeed semi-fixed products, and unlike the fibers in the mitotic apparatuses isolated using the Mazia and Dan protocol, the molecules that make up spindle fibers in living cells do respond to colchicine and cold (and also to hydrostatic pressure, as shown by Ted Salmon during his thesis work in my lab) by reversibly depolymerizing. Conversely, they polymerize further, again reversibly, when the temperature is elevated, hydrostatic pressure is reduced, or when exposed to heavy water (see Fig. 4.5; *Collected Works* A-25, A-27).

It took two decades for people to be satisfied, but I'm glad I stuck to my interpretations of the molecular events, deduced from observations of the birefringence of spindle fibers and filaments in living cells, made with the polarizing microscope. On the other hand, the notion that depolymerizing microtubules could generate a pulling force took even longer to be accepted, in fact, some 40 years as described earlier (see also *Collected Works* A-58).

Chapter 3
Settling in the States

In Post-War Japan, Briefly

Commute to Toritsu

While we made several good friends in Seattle, we were only to stay at the University of Washington for 2 years until 1953. That was because Katy Dan had joined the Biology Department at the newly established Tokyo Metropolitan University as head of the Embryology Section (at the urging of his classmate Daigoro Moriwaki, who had just accepted the chairmanship). He wanted me to return to Tokyo and join him as Assistant Professor in his new group. It was Katy and Jean who had gotten me the opportunity to come to the United States in the first place, and Sylvia agreed that I owed it to them at least to try and see what we could do there so I accepted.

But even as we were preparing to leave Seattle, we ran into our first impediment. Richard Hill, Sylvia's former husband and a merchant marine, went to court to prevent us from taking Heather with us. Even though he himself was out to sea much of the time, he convinced the judge that Heather would be better off with her grandmother in Alderwood Manor because our place in Magome was not fit to bring Heather: "that the road in front of our home was unpaved and that even the house was unpainted." Neither the fact that the Japanese cherished the beauty of natural wood finish, nor the fact that Heather should be with her mother and new family, persuaded the judge otherwise. So Sylvia, 8 months pregnant with our first child, had to sail off to Yokohama on the Hikawa-Maru without her 4-year old daughter. I was to follow a week later after finding a buyer for our car and other belongings.

A month after we arrived in Magome, we rushed Sylvia to St Luke's Hospital in Tsukiji, Tokyo. As my mother and I waited anxiously outside of the delivery room, a young doctor came out to congratulate us for the birth of our healthy, fine daughter. He must have been an inexperienced intern, for an elder physician coming out of the delivery room right behind the first doctor told him, "You'd better be more observant, he's a fine boy!" Thus, on October 5, 1953, our first son, Jonathan Hiroshi

S. Inoué, *Pathways of a Cell Biologist*, Springer Biographies,
DOI 10.1007/978-981-10-0947-1_3

Fig. 3.1 Shinya as assistant professor at Toritsu (Courtesy SI)

Inoué, was born. Hiroshi was a name I chose since it symbolized magnanimity as broad as the ocean.

Life in Tokyo, 8 years after the end of the war, was quite trying for a young Assistant Professor (Fig. 3.1). We were unable to find an apartment of our own; the only one affordable had one small un-insulated room and a narrow hallway with a cold-water tap, with not even a faucet or a sink, let alone a private bathroom or a kitchen, so we stayed at my parents' place in Magome, using the room that used to be mine, and then my youngest sister Futaba's, study. It took half of my salary to buy the condensed milk that was needed for Jon.

Most days I pumped the bicycle, that father had bought for me, for the 40-min-or-so trek to the University. Then, in the introductory biology class, I faced the most hostile students. Was it that they saw me as a lackey of the enemy who had A-bombed, conquered, and occupied Japan? I never did quite understand their hostility, but they may have had the same background as the students who eventually threatened Katy's life after he was elected President of the University during the student uprising in the late 1960s.

Family in Magome

Earlier, father was "purged" by the occupation courts on May 14, 1948, for having held high offices in the pre-war Japanese Foreign Office and subsequently as Civilian Governor of occupied Dutch Borneo (i.e., he was placed on the list of those prevented thereafter to hold any significant public or educational posts). But in October, 1950, he was "de-purged" after his activities were examined in detail. By the time Sylvia and I arrived in Tokyo, he headed KBS (Kokusai Bunka Shinkohkai), a semi-public foundation established for the promotion of international cultural exchange. Mother headed the International Women's Round Table Club, so that both of my parents had returned to activities suited to their personalities and experiences.

It was probably through father's recommendation that, one day, I was asked to come to the Imperial Palace to give a lecture and slide show to Emperor Showa and his laboratory associates. I appeared with my 16-mm time-lapse movies of cell division and must have talked for some three-quarters of an hour on the research that I had been doing in the States. Unlike the experience at Misaki, where there were interested reactions to my live demonstrations, I don't recall what interest those movies and my talk elicited. On the way home, I was given a pack of cigarettes embossed with imperial crests, a memento that used to be given to soldiers in the front lines. Since I did not smoke, I gave the pack to Sylvia, but she doesn't remember much about the flavor of those cigarettes either.

In addition to Sylvia, Jon, and me, my second younger sister, Midori, and her family also occupied another room at my parents' place at Magome in Ohmori, as did Futaba, my unmarried youngest sister. Midori and her family lived on the first floor at the other end of the house from our second floor room, in what used to be the children's room as we grew up. When designing this house after the big Kanto earthquake, grandfather Kiyomatsu had the foresight to cover the floor of that room with cork, unlike the other tatami covered rooms, so that we kids could freely romp about.

Futaba spent many hours practicing her piano in the western-style guest room, also on the first floor below us, which overlooked the pine-tree-clad hillside and the big farm owned by our landlord. She soon studied in France for several months and, as I mentioned before, has become an outstanding concert pianist known for her exquisite sensitivity. She also taught, and became head of the Piano section, at the Elisabeth University of Music in Hiroshima (Fig. 3.2).

Midori had married Setsuo, a composer who was still struggling in those days to make ends meet, and they had a young girl named Fumi, a year or so younger than Heather. Fumi was my father's favorite grandchild. She was very fond of Sylvia, approaching her all the time as her "Sirubia Obachan" (Aunt Sylvia) (Fig. 3.3). Fumi grew up following my parents' interests in promoting international friendship and was to spend her high school years with us in Philadelphia. About the same time that Sylvia had Jon, Midori had her son Kotaro, who has since become a well-known jazz composer and performer (Fig. 3.4).

When Jon was born, the oldest of my younger sisters Sumiko, who had a wartime traditional Japanese marriage to a diligent industrial chemist Shigeru Fujii, already had a daughter Mariko and a son Akira. After his retirement from Mitsubishi Chemicals, Shigeru became an avid artist, winning several prizes at exhibits in Ginza. Sumiko became a master calligrapher of the ancient 'Edo Kantei-Ryu', (Fig. 3.5, *Color version on p. xii.* Sumiko's hand writing in Edo Kantei-Ryu) and still occasionally gives piano lessons at her home.

Fig. 3.2 Futaba solo performance at Hiroshima

Fig. 3.3 Fumi with her Grandmother Hideko in Magome

Fig. 3.4 Kota leading his band

Fig. 3.5 Sumiko Fujii, calligraphy master, and her hand writing in Edo Kantei-Ryu

My third sister, Akiko, who married her chosen classmate at the Ueno Music Academy, Shizuo Horiuchi, also had her first son within weeks of Jon's arrival. Shizuo, a cellist, was eventually to become the concert master for NHK (Nippon Hou-sou Kyo-kai) orchestra. This meant that Mother was busy helping out with all the new arrivals, in addition to entertaining the many visitors and preparing her delicious cakes using the only available foot-cubed small oven perched on her gas stove.

The year in Tokyo was a true trial and, unfortunately, a failure as far as I was concerned. I had not been able to provide Sylvia with much comfort. Jon, while a healthy baby, had little to do except occasionally get into mischief–he insisted on crawling under the old-style washing machine and fingering the oil pan, and he was surprised more than once sticking his fingers into electrical sockets. My undergraduate teaching at Toritsu seemed to be having worse than no impact, and I was getting nowhere in research.

Fortunately, the faculty members at Toritsu were friendly, and Sylvia still remembers the visits with the Moriwakis, our trip to visit Katy and Jean at the marine lab in Misaki, and the visits to our home by Laura and Arthur Colwin who were working with Jean and Katy on the acrosomal reaction at Misaki. Sylvia also accompanied my mother to activities of the International Women's Round Table Club, experiencing classical tea ceremony, the ancient Gagaku performed at the Imperial Palace, and concert by the famed, blind Koto master Michio Miyagi.

Sylvia and I also had pleasant walks among the green fields not far from our home in Magome and also in the hillsides in Karuizawa near Mt. Asama, together with my parents and sisters. Some weekends we went (window) shopping to the Takashimaya Department Store basement where many classical house treasures were on display for sale.

One day, Sylvia and I had a memorable visit with Mrs. Kanokogi, originally from Poland and one of my favorite teachers of German at Musashi. She greeted us in flawless English and, over tea, told us of her recent interest in English poetry. It seems that after mastering some seven different languages, she did not find it difficult to add a new one. To us young parents, who were about to pick up Jon immediately when he had awakened and started crying in his baby carrier, she suggested it was not necessary to jump and cuddle the baby each time he cried.

But we were virtually penniless, and I was getting nowhere in my work, so despite the desires of Katy and Jean that I contribute to the re-building of science in Japan, and my fondness for my family in Tokyo, I told father that I had better return to the States. He was sympathetic, although perhaps disappointed at losing us, and told me that he understood how I wished to make better progress in my research. So I wrote Stan Bennett in Seattle, who responded negatively about my returning to the University of Washington. I also wrote to Kenneth Cooper.

New Life in the States

University of Rochester

Kenneth was Chairman of Biology at the University of Rochester then and had recruited several bright young faculty members including Gus Doerman and Arnold Raven. After letting me know that I should not have written the kind of letter as I had to Stan, Kenneth wrote that he would see what could be done to get me back to the States. Shortly, he and Dr. Morton at the Medical School arranged for a fellowship for me, and we were to join Ken's Biology Department on the River Campus. After 1 year in Tokyo, we sailed back for Portland on another freighter, this time together with Jon.

The trip across the Pacific, my seventh and last ocean-crossing steamer trip, was fairly uneventful, accompanied by the delicious cuisine and good service provided for the few passengers that were aboard the freighter. The only concern was the occasional swaying of the ship, which made Jon fall off the bench by the table, where he insisted on climbing despite his yet unsteady legs. Even though Jon had an exceptionally hard skull, as I had experienced many times when he would butt his head and make my lips bleed while holding him in my arms, the floor of the ship was also solid and his acrobatics were worrisome. But this was not the only occasion that Jon and later his brothers scared us by climbing to unsafe heights.

As we sailed up the Columbia River, what struck me most were the upright and straight trees along the banks of the river. They appeared tall and Lincoln-esque, in contrast to the gnarled pines of the bonsai and the windswept Japanese shore (also seen along the cliffs of Monterey Bay in California). In this vast land of the free, even the trees grew straight and tall, rather than bowing to the pressures of tradition or of the society that constrained the crowded populous of the Japanese islands.

From Portland, Oregon, we drove to Seattle in a rental car arranged for us by Mrs. Anderson, a family friend from our Rose Heights days. Once in Seattle, we were re-united with Heather and visited the Robertsons, Sylvia and Kenneth's family friends, at their home on Bainbridge Island. The picture of Heather, gently taking care of her new young brother Jon on the Robertson's lawn, still remains in my heart (Fig. 3.6). On the shore in front of the Robertsons, we were introduced to geoducks (pronounced gooey-ducks), giant clams which we'd catch by grabbing hold of their foot-long siphons before they escaped by sucking themselves deep into the wet sand. The meat from a couple of the clams (that were so fat that they couldn't fit between their small bivalve shells), provided a delicious meal for all six of us.

In Seattle, we also visited our good friends, the Roosen-Runges. When Edward Roosen-Runge from the Anatomy Department, who had put together his own harpsichord, played his piano for us, we were touched by how deeply absorbed Jon was with the music. Jon displayed a keen sensitivity to music at that early stage which continues to this day.

In Seattle Heather was to stay with her grandmother for two more years while she went to kindergarten and first grade. So our family of three headed for Rochester

Fig. 3.6 Heather with Baby Jon

where a new job was waiting for me in Ken Cooper's department. We don't recall driving there, but neither can we remember a plane ride. In any event we arrived safely and settled into our new apartment at University Park arranged for us by Kenneth and Ruth Cooper.

A day or so after we arrived at University Park, a University of Rochester faculty housing complex, we were called by one of our neighbors saying that young Jon, who could still barely walk, had climbed the painter's ladder up to the second floor window level. He was so full of energy and always wanted to climb! I believe I had to rescue him through the window.

In the apartment across from us lived Isadore and Rhoda Levine from the Med School and their young family. Their boys, and ours of the same age, Jon and Christopher (our second son was born in Rochester on February 5, 1955), would often play together in a wading pool on the lawn protected between the two apartment buildings.

Kenneth Cooper's Department at the University was bustling with energy. Gus Doerman, together with Martha Chase of the Hershey and Chase fame, and their graduate students were exploring the function of DNA whose structure and anticipated genetic role had just been published by Watson and Crick the previous year. Complementing Doerman's group's experiments using bacteriophage, Arnold Raven was using bacterial transformation to probe their genome. Trained by these active and dedicated instructors, the graduate students, such as Frank Stahl and Dick Epstein, were soon to become notable leaders in the evolving field of molecular genetics.

Then, there was Hans Holtfreter, the world-renowned embryologist, who was also quite an artist. In fact, as a youngster in Germany, he apparently had a hard time choosing art or science as a career until he ran into Spemann's influence. Even in Rochester, Hans often appeared lost, saying that modern science had left him behind, until eventually he married his graduate student from Japan, Hiroko Ban. Hiroko brought Hans back to life, and we frequently enjoyed their company and Hans's private art show at their home, also near Landing Road only a few blocks away from our own. Our boys may still remember the clever magic tricks that Hans would perform for them.

Unlike in some departments where the influx of faculty representing new fields of biology created an irreparable schism, Kenneth successfully integrated the new arrivals with those who were already in the department–Babbitt Brown the botanist, Bill Muchmore the entomologist, and Jim Peskin who was interested in the chemistry of vision. It was a friendly and exciting department to join, and I soon felt very much at home. Later, Tom Punnett, Tom Bannister, Jerry and Rachel McMaster Kay, Sandy Segal, and Dick Lewontin also joined us in Kenneth's department.

In the Biology Building, I shared a laboratory with Jakov Krivshenko, whom Kenneth had recruited from the Ukraine. In the Ukraine, Jakov had lost the special hybrid quails that he'd developed, first to the German and then the Soviet soldiers. An accomplished fruit-fly geneticist, cheerful and philosophic, Jakov was a delightful colleague to befriend. One day he prepared microscope slides of special crosses of fruit flies so that Kenneth could verify an important hypothesis of his own regarding a basic issue related to genetic 'crossing over.'

Two months later, Kenneth was still sweating over the many slides that Jakov had prepared for him; the chromosome patterns just made no sense to Kenneth. Finally, Jakov confessed, "You were absolutely right in the first place. I mislabeled the slides for you, so that you could really test your hypothesis." How Kenneth looked at Jakov's action, a strict scientific test and a disguised practical joke, I don't know, but the incident does reflect two aspects of Jakov's characters: desire for irrefutable experimental proof and a philosophical sense of humor.

In a graduate seminar course I was conducting in Rochester, contrary to the assignment that everybody was to read the papers so that they could all join in the discussions, I soon found that only the individual assigned to report on the papers was, in fact, reading them. I told the class that from now on, I would wait until the time of the seminar to say who was going to be reporting today. The next hour, the poor student who had to report was literally quaking at the blackboard, and Margaret Aronson came to plead that at least I let them know the morning of the class who would be reporting today "so that the person would be psychologically prepared, rather than be faced with sudden death."

Actually, that was not a poor group of students at all, and I was only trying to help them become acquainted with a few selected papers that seemed relevant to cell biology. In addition to Margaret were her husband, John, who became my first PhD student, Frank Stahl, soon to become known for the famous experiment of 'Meselson and Stahl' showing the semi-conservative nature of DNA replication, and Dick Epstein, another bright phage geneticist from Gus and Martha's group.

American Optical Company, the Polarization Rectifier

On my part, I was frightened of teaching a course on polarized light in the Institute of Optics (which was part of the Physics Department) where Bob Hopkins was Director. True, I had quite a bit of practical experience, but to teach physics graduate students with my poor background in math! That class, where I literally had to learn much more about polarized light than I had ever known before, lasted for the last 2 years while I was in Rochester. My first encounter with Bob Hopkins was soon after I arrived in Rochester when, at the suggestion of Ken Cooper, Bob introduced me to Brian O'Brien who headed the Research Center of American Optical Company (AO) in Southbridge, Massachusetts.

At our meeting on the campus lawn in front of the Physics Building, Brian O'Brien challenged me by suggesting that AO would sponsor my research if I could improve the extinction factor (ratio of the intensity of light coming through parallel polarizers to that coming through crossed polarizers) of the polarizing microscope by five-fold. That seemed like a tall order, but worth shooting for, so I said I would try. What developed was a delightful interaction with the AO folks at Southbridge and also with the folks at Buffalo where AO had their microscope manufacturing group and friendly individuals such as Bob Tuckaberry.

My first seminar at the AO Research Center in Southbridge, in which I summarized my biological work and the limit of current polarizing microscopes, elicited a most heartening response. In the audience were notable experts on physical optics and microscopy such as Harold Osterberg, Helen Jupnik, Steve McNeal, Oscar Richards, and many others. Among the several suggestions that were made for improving the extinction factor (reducing the depolarization at air-glass interfaces) was one made by Lem Hyde. He said, "Why not place a birefringent half-wave plate between the condenser and objective lenses?" I asked him what that would do and was told that the half-wave plate would switch the direction of rotation of polarized light that was causing the problem, and that the rotations in the condenser and the objective lenses could then cancel each other out. That suggestion flashed in my mind like lightening.

I responded by pointing out that the half-wave plate probably wouldn't work well between the condenser and objective lenses since the high NA light beams were tilted there at high angles. However, if placed at the aperture planes before the condenser or after the objective lens, it might well work if one added a zero-power hollow glass hemisphere with appropriate curvature. I already had data to show that an appropriate number of coverslip surfaces placed in the converging beam between condenser and objective lenses introduced the same amount of rotation as the lenses themselves (*Collected Works* A-8), and that the total effect introduced by the coverslip surfaces could be reproduced by a hollow hemisphere placed at the back aperture of the lenses. Then the zero-power hollow glass (or meniscus) lenses with appropriate curvature would introduce the same rotational distribution as the lenses themselves while the half-wave plate switched the directions of rotation. The

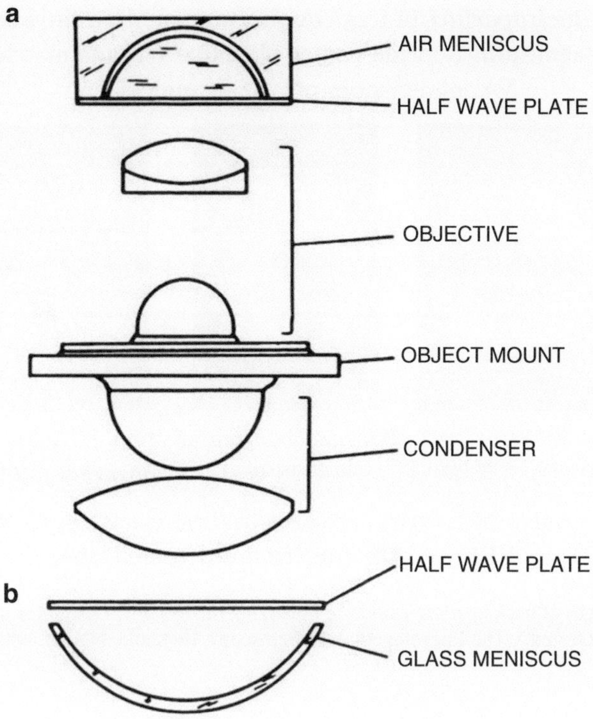

Fig. 3.7 The Polarization Rectifier: The MENISCUS RECTIFIER compensates depolarization of light at air-glass interfaces, thus vastly improving the extinction factor (EF) and eliminates diffraction anomaly. (*Collected Works* A-14; see also Shribak et al. 2002.) (Inoué S, Hyde WL, Studies on depolarization of light at microscope lens surfaces. II. The simultaneous realization of high resolution and high sensitivity with the polarizing microscope, J Biophys Biochem Cytol 3 (6): 831–838, 1957. [A 14])

rotation in the objective or condenser lens and its meniscus lens would then cancel each other out. Thus was born the polarization rectifier (Fig. 3.7).

In a few weeks, I received the meniscus lenses made by the Optics Shop at the Research Center in Southbridge. To each I combined a piece of mica, split to give a half-wave birefringence. By placing these rectifiers into a pair of 97 power 1.25 numerical aperture strain-free oil-immersion objective lenses, which I used in place of the conventional condenser and objective lenses, the extinction factor was now 20 times greater than without the rectifiers installed (Fig. 3.8a)! Even with these high numerical aperture oil-immersion lenses, the image was free from spurious resolution (Fig. 3.8b), and the field was now so dark that the oral epithelial cells, that were virtually invisible before rectification, were now plainly visible, and their birefringent, fine surface ridges were clearly resolved (Fig. 3.8c). So we had met and even exceeded, by four-fold, Brian O'Brien's challenge to improve the extinction factor by a factor of five! (*Collected Works* A-14.)

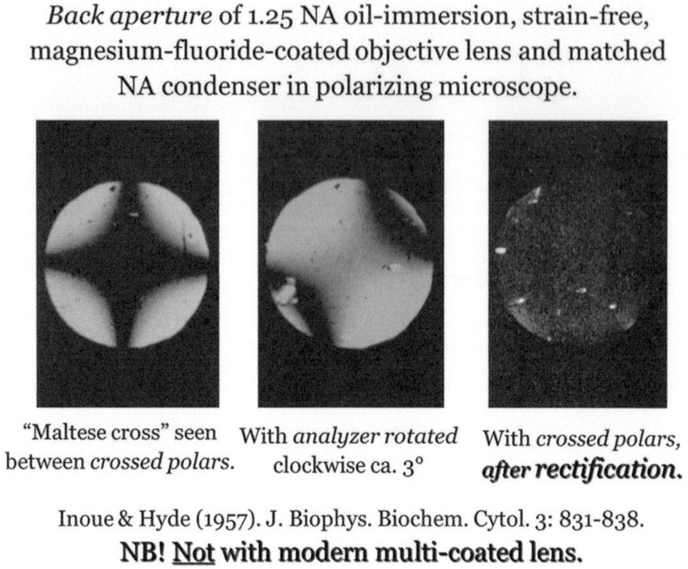

Fig. 3.8a Images of back aperture (Inoué S, Polarizing microscope: Design for maximum sensitivity in Clark GL (ed.), The Encyclopedia of Microscopy, Reinhold, NY, pp. 480–485, 1961. [A 18])

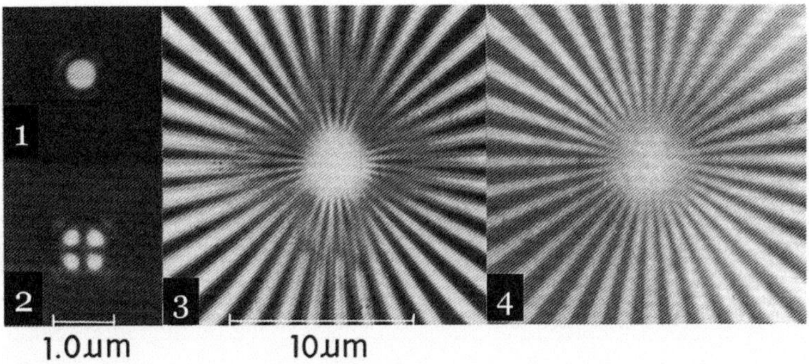

Fig. 3.8b Airy diffraction patterns of pinholes with (*1*), and without (*2*) rectifier; Siemen's test pattern without (*3*), and with (*4*) rectifier, all with 90x/1.25 strain-free objective (Inoué S, Video Microscopy. Plenum Press, New York, 1986. [Appendix V as A 40])

I thoroughly enjoyed my association with the AO group beyond the successful development of the polarization rectifier. Together with Helen Jupnik, we investigated how a large variety of anti-reflection coatings affected the state of polarization at high angles of incidence. (Most of the multi-layered coatings introduced unacceptably high degrees of birefringence.) Harold Osterberg, well known for his mathematical formulation of polarization and related physical optical phenomena, patiently explained their meaning to me in concrete, less mathematical terms that I

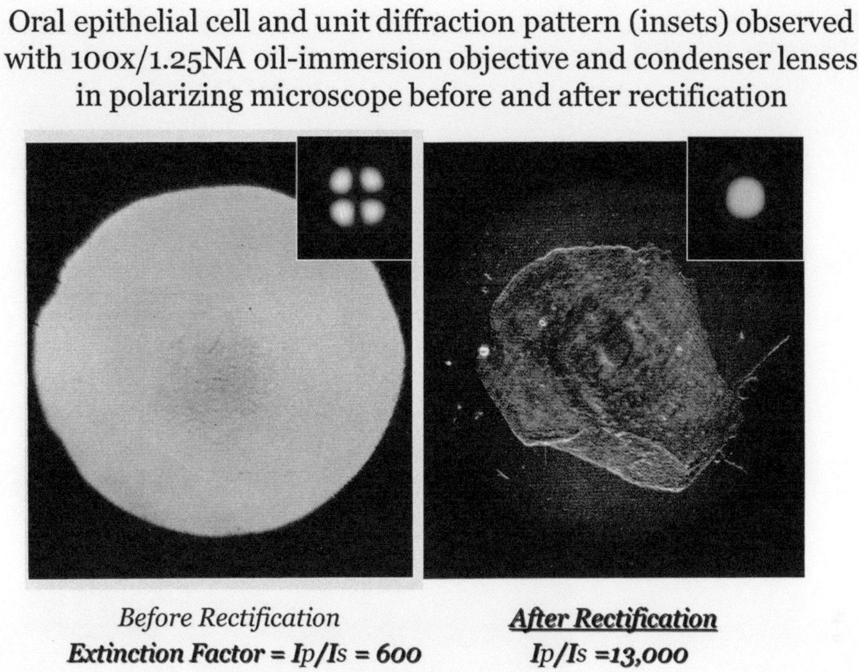

Fig. 3.8c Oral epithelial cell and unit diffraction pattern (*Collected Works* A-14.) (Inoué S, Hyde WL, Studies on depolarization of light at microscope lens surfaces. II. The simultaneous realization of high resolution and high sensitivity with the polarizing microscope, J Biophys Biochem Cytol 3 (6): 831–838, 1957. [A 14])

could understand. I also carried out a number of projects with Chuck Koester, who was initially a graduate student at the Institute of Optics at the University of Rochester and who eventually became Director of the AO Research Center.

In addition to these studies related to optical principles, Warner Engel at the AO Research Center, an astute and accomplished engineer, introduced me to various important aspects of instrument design. Many of his suggestions, and actual components that he designed, were incorporated into Shinya-scope 4 (Fig. 3.9).

For example, as bearings for the polarizer/compensator unit, he fitted precision-finished, tapered-steel plates into the aluminum casting. He told me that the aluminum casting would otherwise deform in a year or two, while the steel plate with the one-inch-broad rim would resist the change and provide the needed bearing surface for a long time. Forty-five years later, those bearings are still allowing us to set the polarizer and compensator orientations to within a thousandth of a degree precision.

While the fourth version of the Shinya-scope was being fabricated at Bob Hopkins's Institute of Optics Shop at the University of Rochester, I had an interesting experience with Rudi Hamburger, a highly skilled instrument maker. He was one of the instrument makers who had come from Germany to Rochester, home not only to the University with its well-known Institute, but to Eastman Kodak, Bausch

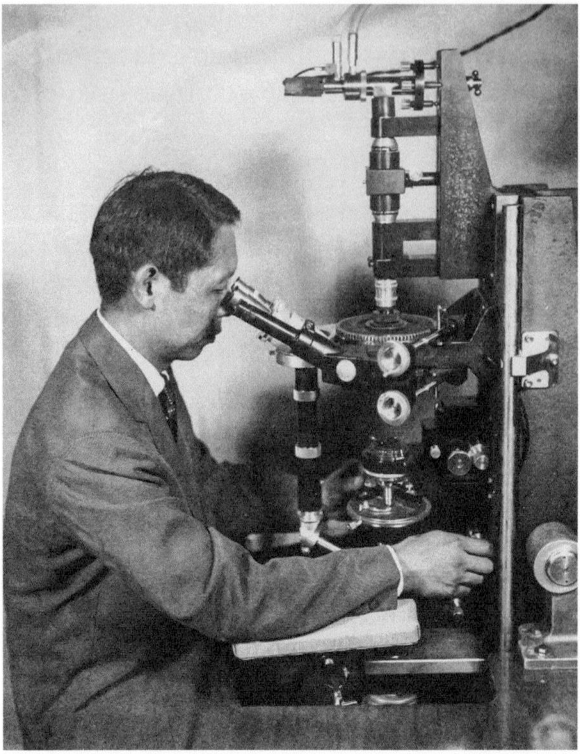

Fig. 3.9 Author using Shinya-scope 4 (Courtesy SI)

and Lomb, Xerox, and other optical companies. I needed a stainless-steel housing for the AH-6 high-pressure mercury arc lamp, that ran on 1000 Volts DC, but which had to be water-cooled and perched on top of my inverted microscope. Well, my design for this lamp involved a short side arm that held a quartz window, with the assumption that the side arm and the upright main housing would be fabricated out of one large piece of stainless steel. Rudi, however, told me how expert they were at welding and so would make the side arm and main body out of two pieces of stainless steel which was much more economical in use of material and fabrication time.

So he made me this beautiful piece, but after only a week it started to drip the cooling water from the joint of the side arm. Fortunately, I had installed a large shield under the lamp, so no harm was done to the microscope components below. Still I had to go downstairs to the shop to see what had been going on. It turned out that the side arm for the window had been fabricated out of a separate stainless steel rod than the main body of the lamp housing. When I placed a piece of paper towel moistened with saliva between the two source rods, and connected the two rods to a volt-ohm meter, we found a significant voltage that developed between the two. Whether that was the source of the electrolytic action that led to the leak, or whether it was more the welding material itself, I really don't know, but the demonstration was enough to convince Rudi to remake the lamp house out of one large piece of stainless steel. The new lamp house he re-made for us (Fig. 3.9) served us perfectly for over 10 years.

While we were working on the polarization rectifier problem, Bob Hopkins introduced me to Dr. Hiroshi Kubota. He was a visiting professor from Japan, an expert in physical optics, who was looking for an interesting project. Since I knew that there was something strange about the image of weakly birefringent, small objects observed in conventional polarizing microscopes, I suggested that it would be worth calculating the unit diffraction patterns in rectified and non-rectified polarizing microscopes. While Dr. Kubota carried out the theoretical calculation, I would pursue the problem empirically.

What we found, through theory and through actual observation of a tiny pinhole, was indeed revealing. Instead of a single diffraction pattern (an Airy disk surrounded by faint symmetric rings), the pinpoint of light through a **non-rectified objective produced a four-leaf clover pattern in the image and the fine diffraction pattern was anomalous** (Fig. 3.8b). In other words, with a conventional, non-rectified polarizing microscope, each point in the specimen became represented by four closely-spaced points in the image. That finally explained why fine specimen details, such as the band pattern in cross striated muscle, could be misleading; the birefringent regions could appear non-birefringent and the non-birefringent adjacent regions could appear birefringent depending on their spacing. By using rectified optics, the diffraction pattern was corrected, so that one could now follow the birefringence distribution without any image anomaly (*Collected Works* A-15, A-17).

One interesting character in Bob Hopkins's Institute was Ted Dunham who had hailed from Australia. He was an astronomer who proposed to introduce the astronomer's technology to biology, and had set up one of the few existing Birch microscopes. This was an elegant looking instrument equipped with special optics, a massive stand whose design was based on fully kinematic support, and housed in a glass cage for temperature control.

When I visited Ted's lab, he proudly showed me the image of a small pinhole through his Birch microscope and pointed out how perfect the Airy disk diffraction pattern was. I looked through the ocular and sure enough, after about a minute or two, saw a perfect Airy disk. But before the disk appeared, what I saw was a thin line, the width of the Airy disk, which gradually became shorter and shorter until it finally turned into the definitive disk. What happened was that, because the solid mechanical components were all assembled based on strict kinematic support, the whole microscope acted as a tuning fork, which would ring in response to any external vibration. The lesson I quietly took home was that in designing a microscope one should use semi-kinematic rather than fully-kinematic design and that one should incorporate some compliant components which would absorb the vibration.

Andrew and Wishia Bajer and Haemanthus Cell Division

While I was working on problems of microscope design and the basic optics with the folks at AO and with Dr. Kubota, we had a memorable visit from Andrew and Wishia Bajer (pronounced Buyer) who had recently arrived from Krakow, Poland, to Canada. The Bajers had been studying chromosome movement and cell division

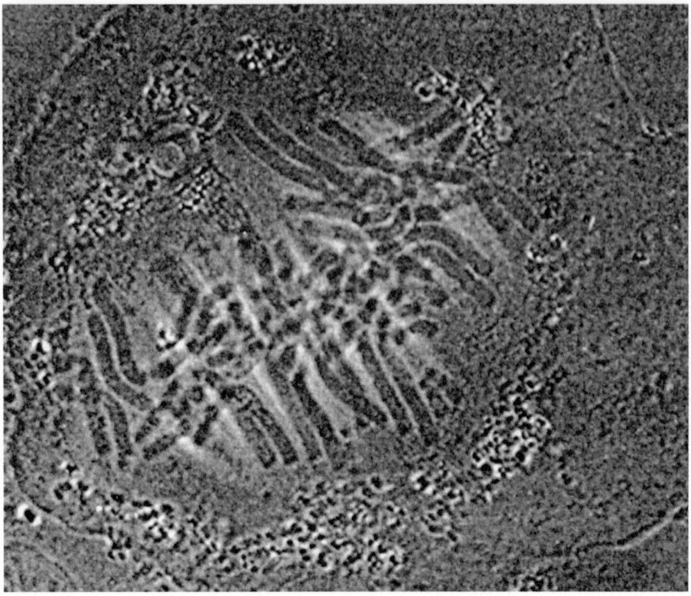

Fig. 3.10a *Haemanthus* endosperm cell in early anaphase (Inoué S, Organization and function of the mitotic spindle in Allen RD, Kamiya N (eds.), Primitive Motile Systems in Cell Biology, Academic Press, New York, NY, pp. 549–598, 1964. [A 22])

in plant cells and had produced some fabulous time-lapsed movies using phase contrast microscopy.

Shortly before arriving in Rochester, Andrew Bajer called and asked me to get hold of some flowering *Haemanthus* (African blood lily) plants. Upon learning from me that we could not locate any plants here, Andrew was indignant, "What, in a rich country like America, there's no *Haemanthus!*" Such was a typical outburst from impatient Andrew. He was the one who had dragged his sedate Swedish friend Gunnar Oestergren for many feet on the pavement, not realizing that Gunnar's coat was caught in the door of the Volkswagen that Andrew was driving.

So Andrew and Wishia came to Rochester bringing a few pots of *Haemanthus* plants procured in Canada, with flowers at the right stage. Patient Wishia was an expert at making the preparation of endosperm cells spread on a sandwich of agar, so we were soon observing the nicely flattened cells under my polarizing microscope. There, we could clearly see the chromosomes as well as the spindle fibers that led the chromosomes to the spindle poles (Fig. 3.10a). As well, we could see the new fibrils of the phragmoplast that appeared between the separating chromosomes and accumulated the vesicles, which would fuse to form the cell plate (Fig. 3.10b). Even though the cellular machinery that moved and positioned the cell organelles could not be seen at all with phase contrast or by DIC (Fig. 3.11), they became clearly visible in polarized light by their birefringence. Now we were getting an even better handle on what was going on in the cell as it was dividing (*Collected Works* A-19).

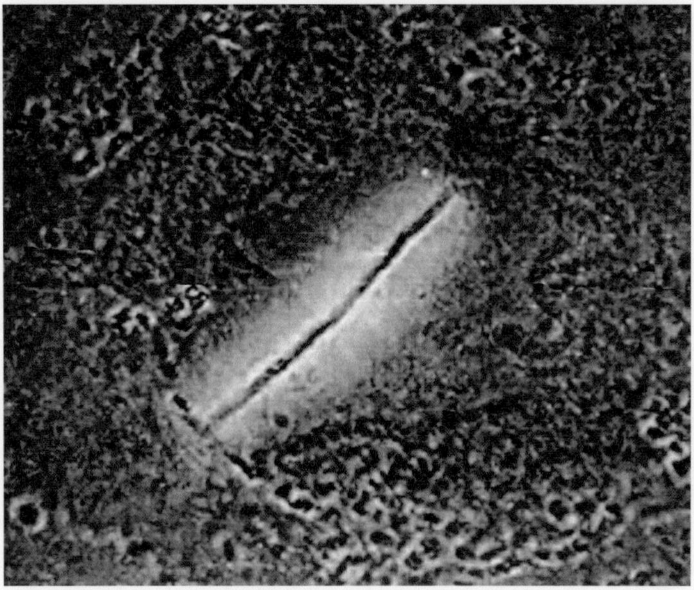

Fig. 3.10b *Haemanthus* endosperm cell at cell plate formation (Inoué S, Organization and function of the mitotic spindle in Allen RD, Kamiya N (eds.), Primitive Motile Systems in Cell Biology, Academic Press, New York, NY, pp. 549–598, 1964. [A 22])

In the meanwhile, the rectified microscope was now completed and installed in my lab, but after all that went into its development, I was too scared to put the system through a real test using specimen besides the optical test targets. Of course, what was really needed was to try out the system on significant biological specimens. It was not until we arrived at Dartmouth in 1959 that the instrument got its real work out, using a specimen suggested by Ken Cooper. But that is still some time off in the future.

Friday Harbor Again

Returning to the early days in Rochester, we decided to spend the summer of 1956 at the Friday Harbor Labs, where we had enjoyed our previous visits from Seattle. That would also give us a chance to visit Sylvia's folks in Walla Walla, Washington. This was our first long cross-country ride in our Ford station wagon with Sylvia just learning to drive. With Jon and Chris still in diapers, our motel stops were gauged by the availability of Laundromats. We managed to arrive safely and as Edward Roosen-Runge from the Anatomy Department in Seattle who was spending the summer there.

At Friday Harbor, I started taking a series of still photographs of the developing jelly fish, *Halistaura* (Fig. 3.12), *Aequorea*, and *Phialidium*, using a temperature-

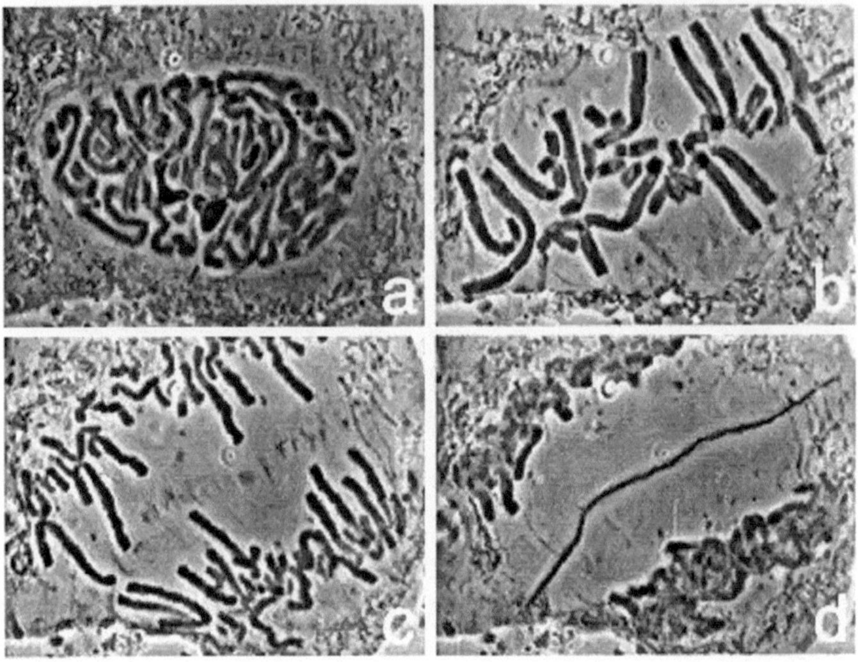

Fig. 3.11 *Haemanthus* endosperm cell: Late-prophase through cell plate formation, observed with DIC microscopy (Inoué S, Oldenbourg R, Microtubule dynamics in mitotic spindle displayed by polarized light microscopy, Mol Biol Cell 9 (7): 1603–1607 (video essay), 1998. [A 62])

control slide perfused with the Puget Sound seawater which remained at 11° C (Figs. 3.13a and 3.13b). I was also able to record the early development of the glass-clear, bullet-shaped jelly fish, *Aglantha digitale*, which swam so fast that few live specimen had ever been caught before.

The day-old *Aglantha* larva was a voracious fellow, opening a mouth wider than its body and swallowing up the drop of blood that spilled into the dish when I accidentally cut my fingertip with a piece of broken glass tubing. That chance observation helped Edward Roosen-Runge raise his jelly fish larvae with no further trouble; they only needed to be fed fresh blood (or maybe it was the glutathione in the fresh blood that they needed)!

Scooping up the plankton that were flowing by the floating dock with a finger bowl, we could collect intact live specimen that would otherwise be broken up if collected with a plankton net by towing. Among the fascinating creatures that I collected were *Tomopterus*, a pelagic polychaete worm that was absolutely glass clear. In fact, besides its eyes and gonads, there was no pigmentation or light-scattering structure in the living worm so that, under a dissecting microscope illuminated obliquely, one could see the small nucleolinus inside the nucleolus within the nucleus of the oocytes all within the intact parapodia.

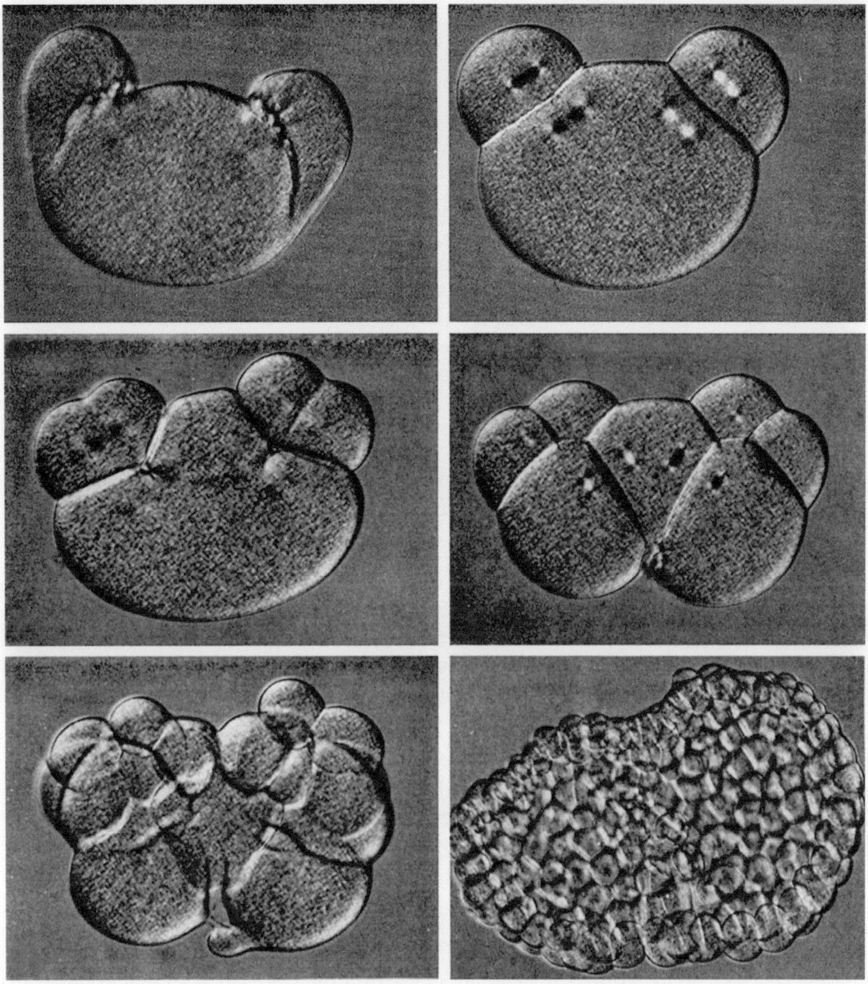

Fig. 3.12 Developing egg of jelly fish, *Halistaura cellularia*, viewed in polarized light. The first division of this particular embryo was prevented from completion by chilling the egg and depolymerizing its spindle (*Collected Works* A-22) (Inoué S, Organization and function of the mitotic spindle in Allen RD, Kamiya N (eds.), Primitive Motile Systems in Cell Biology, Academic Press, New York, NY, pp. 549–598, 1964. [A 22])

Never had I come across such a crystal-clear organism, and I yearned to study its cells, especially since it was a reasonably close relative of *Chaetopterus* whose eggs were so opaque that their spindle was invisible without centrifugation. Unfortunately, the only three specimen that we caught that summer were all females which survived just long enough to tantalize me but not to allow me to study them any further.

In the meanwhile, our children were enjoying the wild rabbits that teased them around the cabins, and swimming in the quarry pond (the water in the Sound was

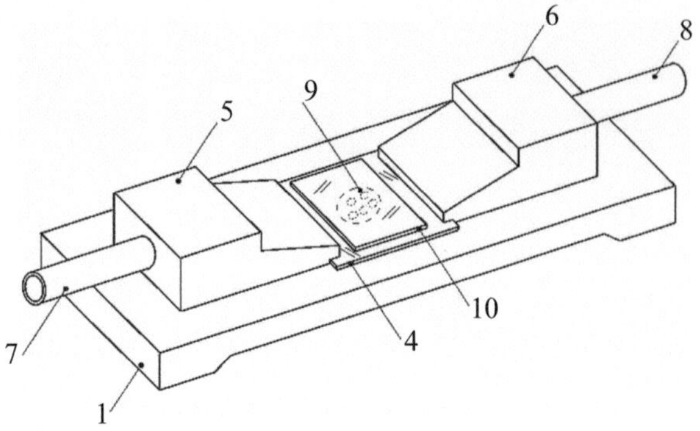

Fig. 3.13a Constant temperature slide overview (Courtesy SI)

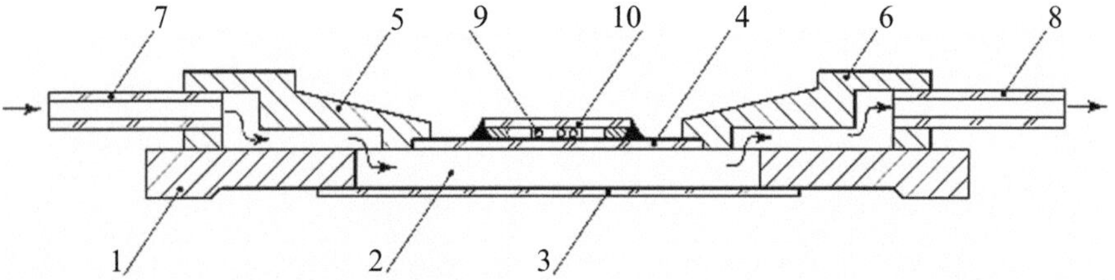

Fig. 3.13b Constant temperature slide cross-section view. Metal base (*1*), Flow chamber (*2*), Coverslips on flow chamber (*3*, *4*), Plastic in-outlet (*5*, *6*), in-out flow tube (*7*, *8*), specimen (*9*), top coverslip (*10*) (Courtesy SI)

much too cold, being at a constant 11° C). Driving to the north of the island, we saw large pods of killer whales, whose water-slapping activities were sometimes mistaken as gun shots. Then there was this enormous tidal flat, False Bay, where at low tide the whole family would join Dixy Ray and me as we dug for *Mesochaetopterus*. Unless we managed very quickly, those worms would slide down away from us into their yard-long, straight parchment tubes.

The 20-foot-long, hollow brown stalks of the Puget Sound kelp that washed up on the beach made kelp horns that entertained us all, especially Heather, with their bassoon-like low resonance. Then there was the salmon bake, where everybody from the whole Friday Harbor Island would congregate and feast on the freshly baked salmon and steamed mussels from the Sound. I can't forget the most delicious shrimp that I had ever tasted in my life; that was on board the Lab's collecting ship, where the captain boiled the shrimp, just netted from the Sound, to perfection.

A new hearing at the end of summer gave us custody of Heather and she joined us on our return trip to Rochester.

In the cramped two-bedroom apartment, Heather had the small bedroom and the two boys occupied the larger bedroom with two maple bunk beds that were arranged separately, as neither of the boys were older enough to sleep on a 'top bunk'. Sylvia and I made do on a hide-a-bed sofa in the living room. When Steve, our third son, was to arrive on May 27, 1957, we would have had to place his crib in the apartment's kitchenette. So by then it was time to look for our own first house, which was partly made possible, Sylvia reminds me, because by then we had some extra income from my consulting work with Xerox and with American Optical Co.

We did find a two-story Dutch Colonial house with enough room for our family in the suburbs of Rochester, a 30-min drive from the University and a five-minute walk to the grammar school for the children, at the end of Landing Road. There we had four bedrooms upstairs, a spacious kitchen and dining room in one half of downstairs, and a large living room with an enclosed porch in the other half. Down below was a full basement with a built-in garage. The day after we moved in, we found Jon, together with Chris this time, again on the ledge of the second floor, three levels above the lawn in the back yard. Why did they need to explore their new residences by giving us such a scare each time.

Biophysical Sciences Meeting

The summer of 1958 we had another trip out West, from Rochester to Boulder, Colorado, this time with another new arrival, our third son Stephen Kenneth, born May 27, 1957. The occasion for the Boulder trip was an experiment set up by F. O. Schmitt, Larry Oncley, and others to bring together not just the scientists who were likely to be early contributors to the developing field of biophysics, but also their families onto one campus (Fig. 3.14). As recorded in the resulting publication, *Biophysical Sciences–a Study Program (1959)*, there was much active discussion and interaction among the scientists, many with different backgrounds, who were brought together for the first time on this occasion. As to the NIH-supported experiment of bringing the families together as well as the scientists, probably that plan was not considered sufficiently successful since, as far as I am aware, the scheme was never tried again.

At the opening session, held on the evening before the scientific presentations were to take place, F. O. Schmitt was so excited about the upcoming discussions that he proposed, "Let's start at 8:00 tomorrow morning instead of at 9:00 as planned," to which the auditorium became dead silent for a minute until Astbury from Leeds stood up and exclaimed, "What, such uncivilized hour!" Fortunately for the rest of us, faced with this exclamation by the pioneer in X-ray diffraction analysis of biological fibers (whom Schmitt himself had invited from England), Schmitt had to back down.

Speaking of Astbury, based on his own X-ray diffraction studies that showed a 3.5 Angstrom periodicity along the length of the DNA fiber, coupled with the negative birefringence found by W. J. Schmidt using the polarizing microscope, Astbury had proposed that the DNA molecule was made of sub-unit molecules arranged like a stack of coins. That was almost a decade before Watson and Crick came up with their double helical model of DNA. The only problem was that Astbury had sug-

Fig. 3.14 Scientists from several disciplines gathered for Biophysics Conference at University of Colorado, Boulder campus, summer, 1958 (Courtesy U of Colorado, Boulder. 1958)

gested four nucleotides per stack rather than the paired two that was essential for the Watson and Crick model.

Despite F. O. Schmitt's eagerness, the whole Conference was carried out at a sufficiently leisurely pace that the participants had plenty of time to become acquainted and interact with each other. One special help for me came from Manual Morales, who enlightened me on the thermodynamics of the equilibrium between the monomer and polymer of the spindle fiber material. During my presentation, I had shown a graph on how the birefringence of the *Chaetopterus* spindle fibers at metaphase equilibrium increases reversibly (to a maximum asymptotic value) with moderate rise of temperature. And I had asked the audience to volunteer with an interpretation of this finding in terms of polymerization of molecules.

First, Manual and I agreed to assume a very simple model: that the concentration (A_0) of the total amount of the material that could make up the spindle was constant whether or not it was already in the spindle. And that, of this amount, some of the material (B) becomes polymerized to form the birefringent fibers, thus reducing the concentration of the still polymerizable material in the pool to (A_0–B).

That gave us the working equation by which the birefringence versus the reciprocal of absolute temperature (1/T) data could be converted to a van't Hoff plot: {ln (B/(A_0–B)} vs. (1/T) (Fig. 3.15). The data, in fact, plotted as a nice straight line from whose slope and intercept, Manual calculated for me the huge increase in entropy and enthalpy associated with the temperature rise. In other words, the polymerization of the spindle fiber material is somehow associated with a huge increase in entropy and enthalpy, while reflecting a very small free energy change (less than 1 kcal/mol at 20° C).

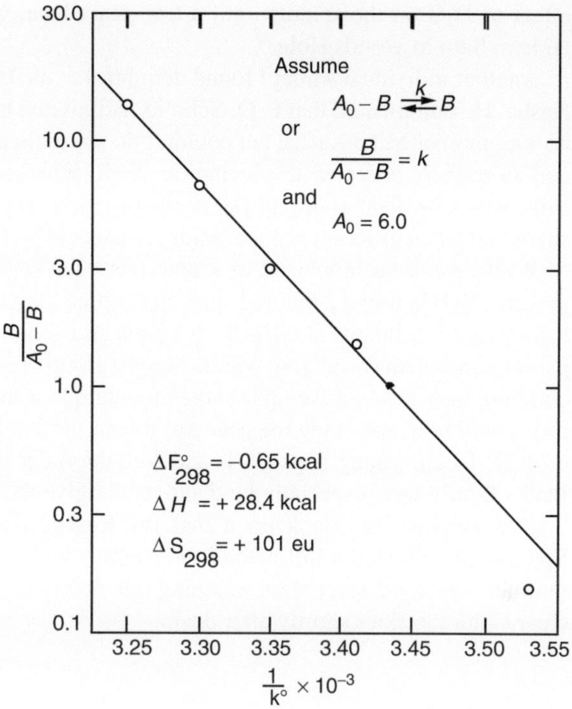

Fig. 3.15 van't Hoff plot of *Chaetopterus* spindle birefringence (Inoué S, Organization and function of the mitotic spindle in Allen RD, Kamiya N (eds.), Primitive Motile Systems in Cell Biology, Academic Press, New York, NY, pp. 549–598, 1964. [A 22])

These data told us that the polymerization of the spindle material was endothermic and entropy-driven. Even though the birefringent polymer was obviously more ordered, the entropy *of the system* was actually increasing by quite a bit as more polymer formed. Taking into account a similar finding by Hans Laufer on the assembly of A proteins into Tobacco Mosaic Virus particles, and Oosawa et al. on G-actin to F-actin transformation, it appeared that while the protein molecules were coming together in alignment (hence the increased birefringence), water molecules, which had previously been bound and ordered on the monomers, were coming off the monomers and becoming randomized; hence, the increase in entropy of the system as a whole. Up to a point, the higher the temperature, the more water was randomized, and the released monomer became pushed out of the aqueous medium (as oil would separate from water); in other words, the monomers stripped of bound water interacted hydrophobically to form the birefringent polymer. (*Collected Works* A-16; see also Salmon 1975, on further thermodynamic studies using the other important state variable, hydrostatic pressure.)

Hearing these discussions, Melvin Calvin of the photosynthesis fame suggested that we try using heavy water. To my question as to what happened to cell division in algae, he answered, "They die," but added that if we were to pursue this, we should get pure D_2O made by BioRad. So, soon after returning to Rochester, I bought a bottle of BioRad D_2O. But, in fact, I did not have an occasion to test the effect of D_2O on the spindle until a few years later, when I tried it together with Hidemi Sato in Woods Hole.

Another individual whom I found delightful at the Boulder conference was Mike Kasha. He commented that F. O. Schmitt had invited him by mistake, thinking that he was another Mike Kasha, but couldn't do anything about the mistake since Mike had so eagerly accepted the invitation. Well, whether or not invited by mistake, Mike was a brilliant young physical chemist, who showed how electrons could be shared between different groups inside a molecule and travel as "Excitons."

While we were immersed in science, our wives and family sought their own pastime. Sylvia found a kindred spirit in Cynthia Davies from England, whose husband was a biochemist at NIH. Both Sylvia and Cynthia preferred their quiet companies rather than socializing with the larger group of academic wives. So with their children, they would drive up into the mountain to a shallow stream where the children could play and enjoy the peaceful forest, far from the dormitory and the laundromat. To our young boys Chris and Jon, the older boys from the physicist John Platt's family were especially kind and would give them fun-filled shoulder rides.

One evening, having learned that the Russians had launched a satellite–the Sputnik–we climbed a hill behind the campus with the Astburys and a few other families. We were speechless watching the slowly blinking 'star' travel across the clear Colorado skies shortly after dusk.

Another Move

Hanover, New Hampshire

In 1959, the summer after the Boulder meetings, we must have driven towards Woods Hole from Rochester. I distinctly remember the long New York State highways, traveling for miles along the Eerie Canal north of Syracuse, and past the sign for Herkimer diamonds. We must have had four children plus some pets in the station wagon, but neither Sylvia nor I can remember what transpired in Woods Hole that summer. Perhaps that was because we were not actually heading to Woods Hole, but had turned north after passing the Berkshire Hills in Massachusetts, to go house hunting in Hanover, New Hampshire.

At the February meeting of the Biophysical Society in Boston earlier that year, Manual Morales, whom I had worked with at the Boulder Meeting, had invited Art Samuels and me to visit Hanover to take a look at what was going on at the Dartmouth Medical School. At Hanover, buried in several feet of snow, Manual and the young dean, Marsh Tenney, told me about the Medical School's history and the need to strengthen the basic medical science departments and graduate programs. In brief, Dartmouth had long been recognized as one of the outstanding 2-year medical schools, with its graduates all transferring to Harvard and other prestigious 4-year schools to finish their clinical training. The American Association of Medical

Colleges had decided that the School would lose its accreditation unless the basic science departments were modernized and strong research programs were introduced.

Manual was thus brought in as Chairman of the Biochemistry Department, and they were negotiating for a prestigious individual to head the Microbiology Department. They wanted me to come and build up the Department of Anatomy and Cytology together with a strong graduate program. A new building, to be erected very shortly, was in the design stage and I was free to specify the plans for our department space.

We were very happy in Rochester, and building up a new medical school department and graduate program sounded like quite a challenge. But after all, I had started my career in Stan Bennett's Anatomy Department in Seattle, and from 1955 to 1958 in Rochester I was 'Scholar in Cancer Research' supported by the American Cancer Society, so why not take on the new challenge at the Dartmouth Medical School, even though my training was really in biology?

In retrospect, however, I may not have been as cautious as Katy used to characterize me. He used to tease my hesitance in taking gambles by using the old Japanese expression, "Probing with the walking stick while crossing a stone bridge." When I talked with a senior member of the Physiology Department at the Medical School in Rochester where Marsh Tenney had held a position earlier, I was told, "You sound like you've already made up your mind to go," and so missed out on whatever advice he might have given me had I asked before deciding to go to Dartmouth.

House hunting around Hanover was unlike any other we had experienced. Outside of the quaint college town of Hanover and the sparsely scattered surrounding villages and towns, there were wide open fields and hills that whetted our appetites. The first farm house of interest on a twenty-acre hill side in Vermont was vetoed by Manual's mother who insisted the place was much too run down even to repair. Somehow, we were introduced to the Dailey's who owned the Shipman place in Plainfield, some 12 miles south of Hanover. This old New England estate with many bedrooms, a huge fire-place in the living room papered with tea cloth, and even a 30-foot-long music room with a secret revolving bookcase hiding the staircase up to a secluded bedroom, also had a large barn and a box-wood-encircled, lawn-covered yard. It even had its own brook leading up to a pond and acres of pine woods. It seemed an excellent place for our kids to spread their wings and experience country living, so we rented the place with little hesitation.

The school house in Plainfield had four classrooms with two grades in a room. It was not fancy but adequate, and despite one episode of cooties caught from the coatroom, Heather and Jon were exposed to good teachers. Chris attended Mrs. Perry's kindergarten, up the hill in front of the town school house. At home, the children fed the turkeys and chicken and collected eggs from the hay stack in the barn, chased the bunnies that multiplied over winter in the wood pile under the house. We had two horses, but Perry was too big for the small boys to ride and Albert was a young horse in need of training. Both horses would go through a weak spot in the fence and Sylvia would be called by a neighboring farmer, "Mrs. Inoué,

Fig. 3.16 Chris above Shipman pond checking for fish (Courtesy SI)

your horses are in our field again." Often that would happen at the crack of dawn on a frosty winter morning.

One day 5-year-old Chris came from the pond above our brook and asked for a pin so that he could catch a trout. We didn't have fish hooks, so I gave him a pin from a new cotton shirt and bet him a dollar that there was no way he was going to catch a trout with such a pin that had no barb. Unfazed, he bent the pin into a U, tied it to a string and thin branch, and marched off to the pond again. No bites that evening, but early next morning he came running back excited, "I've got a fish; I've caught a trout!" Somewhat skeptical, I went up to the pond with him, where what did we see but his fishing rod scooting around on the surface of the pond! Sure enough it was being pulled around by a 6- or 7-inch-long trout that had taken Chris's bait, so he won his bet and the $1. Chris continues his fishing enthusiasm in New Mexico and on the Cape to this day, tying his own flies from bird feathers that he and others collect (Fig. 3.16).

While life at the Shipman place was wonderful in many ways, I was having trouble with my return drive late in the evening from Hanover. On the road running straight through a swampy, flat section, the mist became so bad in the winter that I could barely see one segment of the yellow-dashed dividing lines some 15 feet long. In addition to having to inch along the curved road, driving almost blind in the mist upset my stomach; what a miserable way to arrive back home after work!

Fig. 3.17a Sylvia with Baby Ted at Mink Brook

So after a couple of years, we decided to leave Plainfield and look for a place closer to my work in Hanover. Just south of town, a new development was opening up, and we chose a place at the end of Clafflin Circle, with the back of the lot sloping down to Mink Brook. The College arranged a generous building loan, and our home was built following our own design.

With few houses still in the development, the wide open fields were ideal for the kids to shoot off water- or gun-powder-powered rockets, the seldom traveled driveway was safe for little Stephen to dash off on his tricycle, and Mink Brook provided hours of fun for Jon and Chris. Heather was approaching high school age and made many friendships with the children of the Dartmouth faculty. During the winter, the whole family took advantage of the golf course on the other side of town, that was converted to a beginners' ski slope, and of Occom Pond which became the town's ice skating rink. On weekends during the warmer weather, the family would enjoy their outing to the neighboring river (Figs. 3.17a and 3.17b).

Ted, our youngest, was born on July 6, 1964, at the Mary Hitchcock Hospital in Hanover and, in several months, entertained the whole family by scooting around in his baby hopper (Fig. 3.18a). He was such a relaxed baby that he'd often fall asleep on his high chair, ending up nodding with his face in the supper dish.

Steve from his younger days was crazy about wheels (Fig. 3.18b). This foreshadowed his success as a motorcycle racer and eventually the owner of Santa Fe's most-trusted Toyota repair shop (Toy Auto Man).

Fig. 3.17b Jon and Steve at Mink Brook

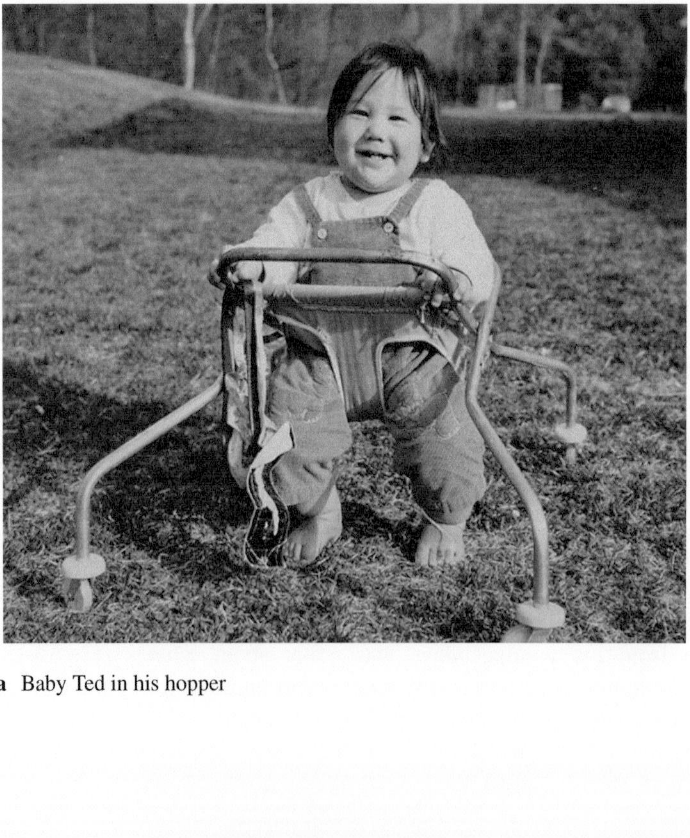

Fig. 3.18a Baby Ted in his hopper

Fig. 3.18b Steve struggling to fix flat tire

Dartmouth Medical School

At the Medical School, my first recruitment was brilliant Wayne Thornburg (Fig. 2.20) who was still at the University of Washington. Together with him, I discussed the makeup of an ideal modern Department of Anatomy and Cytology, as well as design of the space to house our new department. Fortunately, Harry Savage, who had been teaching gross anatomy with help from the physicians at the Mary Hitchcock Hospital, was happy to continue taking care of that aspect of the teaching, so it was up to us to consider primarily the Histology course and graduate education.

For the next recruitment for our department, Wayne and my first choice was Kenneth Cooper (Fig. 3.19), who was then a Research Professor at the University of Florida in Gainesville. With his far-reaching insight into genetics and biology in general, buttressed with what I had experienced assisting him in the Cell Biology Course at Princeton, and as a member of the highly successful Department he established at Rochester, Ken was exactly what we needed for our forward-looking Department.

When I traveled to Tokyo to visit my parents shortly after arriving in Hanover, I interviewed Hidemi Sato (Fig. 3.20), who was recommended by Hiroko Ban Holtfreter, as a candidate for my lab assistant. Hiroko, our friend from Rochester, had known Hidemi's wife Yuki as a classmate and was most impressed by Hidemi.

He had studied the mating behavior of *Paramecium* and had shown that if not allowed to mate with *Paramecia* from another strain for about 100 generations, they would eventually die after a couple of hundred more generations of rescue attempts through intra-strain mating. He had also discovered that the mitochondria in *Paramecia* were replicated synchronously by binary fission preceding the ciliate's cell division. With my nascent interest in protozoology, I immediately offered him

Fig. 3.19 Kenneth Cooper

Fig. 3.20 Hidemi with Yuki and Masahiko Sato (*right to left*), pose next to Benjamin Franklin's statue in University of Pennsylvania campus

the assistantship. Their arrival to Hanover was, however, somewhat complicated. Their younger, 1-year old son, Masahiko, had to join his family a year later, flying all by himself across the Pacific after recovering his health under his grandparents' care.

Sperm DNA

While continuing to culture the variety of protozoa as I had requested, Hidemi and Yuki were also interested in participating in my own on-going research, so after finishing our histology teaching and my chairman's duty during the daytime, we would get together at 5:00 pm and start on the project. The project we tackled was related to the UV microbeam irradiation experiment on chromosomes that Dr. Bloom discussed during F. O. Schmitt's course at MIT almost 10 years before.

One morning, before 5:00 am, Sylvia helped out by driving down to White River Junction to fetch Dr. Bloom, who had arrived by train. He had come to join us in seeing what effect polarizing the UV microbeam would have on metaphase chromosomes of tissue-cultured cells. As I had more or less expected, those metaphase chromosomes (which have virtually no birefringence to start with) showed no differential response to the polarization direction of the irradiating UV microbeam.

After trying hard for 2 days, I suggested we try chromatin in the sperm head instead of metaphase chromosomes. We knew (since W. J. Schmidt's days) that sperm chromatin was highly ordered and exhibited a prominent negative birefringence, so there was a better chance of seeing anisotropic effects induced by polarized UV irradiation. Unfortunately, Dr. Bloom was only interested in somatic chromosomes and not at all in sperm chromatin, so he left before we had a chance to see what would happen to the sperm head irradiated with polarized UV. Dr. Bloom did, however, thank Sylvia for picking him up so early in the morning by presenting her with an autographed solid cherry cutting board that he made out of his precious stock of hardwood kept in his old home in Vermont. This cutting board is still treasured in our kitchen after over 60 years.

By the time Dr. Bloom visited us, Hidemi Sato's excellent talent was recognized by all in the Department and he had unanimously been promoted from research assistant through research associate to assistant professor. We had also hired Cynthia Knudson to prepare the microscope slides for the histology course, and through her, her talented husband Bob. Bob Knudson had been working in a small repair shop in Lebanon, New Hampshire, and was happy to design and set up his own modest instrument shop in our Department. Incorporating new features designed by Gordon Ellis, it was Bob who built Shinya-scope 5 for us. We were also joined by Dick Markley, a skilled and conscientious photographer recommended to us by the Ruebs who owned the Hanover Photoshop. Shortly Bertha Woodward, who was a one-person mail carrier in Plainfield and who had nurse's training in addition to running a dairy farm and rental housing with her husband Ralph, came to take over Cynthia's job. As we will see later, our friendship and support by these loyal New Englanders lasted for over 50 years.

Not long after Dr. Bloom's visit, we tried irradiating live sperm head of the cave cricket, *Ceuthophilus nigricans*, which Kenneth Cooper had collected for us. Those hunch-backed, distant relatives of mole crickets, lived in moist pockets under rocks. Kenneth chose the cave crickets for us following the dictum (from Franz and Sally-Hughes Schrader) that *unusual organisms tend to provide extraordinary gifts of Nature.* Sure enough, the cave cricket sperm glided forward, not propelled by a whip-like flagellum as with most sperm, but by waving an undulating membrane similar to those found on toad sperm. The birefringence of the mature, needle-shaped sperm head (that contains the chromatin) showed a spectacular zigzag birefringence pattern, similar to what one sees in immature grasshopper sperm head.

For the UV microbeam experiment, we made some modifications to Shinya-scope 5 (Fig. 3.21) with Bob Knudson's (Fig. 3.22) help. Since the cement used in the Glan-Thompson polarizer would not transmit the short wave length UV needed to irradiate the chromatin, we placed a tiny first surface mirror (the micro-mirror) below the Glan-Thompson prism, just in front of the field diaphragm. The micro-mirror, which acted as the miniature source of UV, was tilted at 45° to the microscope axis and illuminated by a second mercury arc lamp. That lamp, equipped with

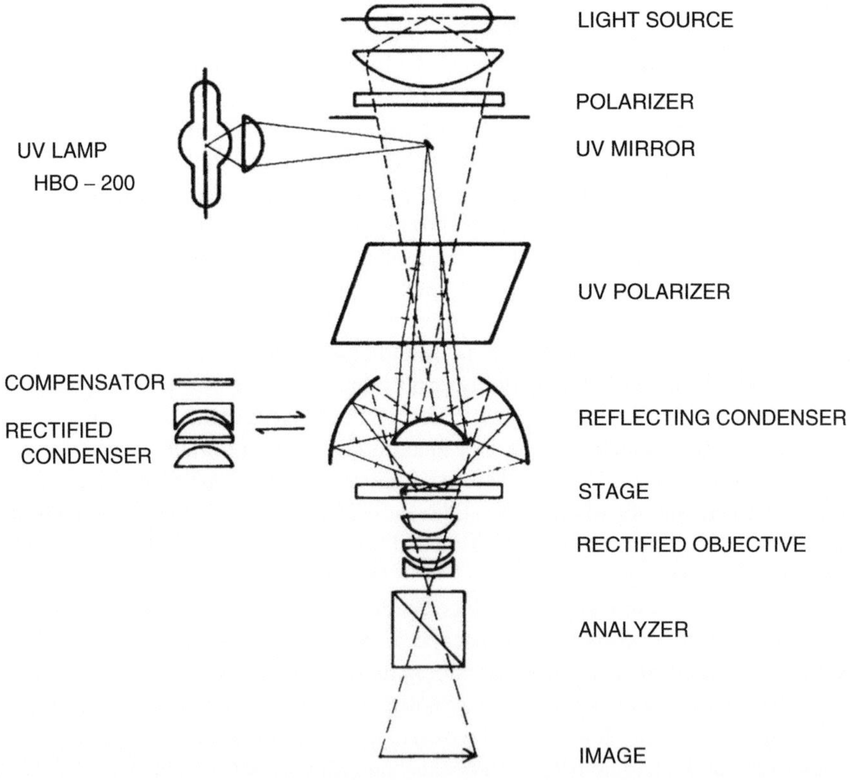

Fig. 3.21 Schematic of modified Shinya-scope 5, incorporating polarized UV microbeam system (Inoué S, Video Microscopy. Plenum Press, New York, 1986. [Appendix V as A 40])

Fig. 3.22 Bob Knudson (Courtesy SI)

quartz collecting lens and an electric shutter, was mounted on the side rail of the tall inverted microscope.

For UV micro-beaming the specimen (that we mounted between quartz cover-slips), we used a 100x Zeiss Ultrafluar objective lens in place of the regular condenser. The Ultrafluar objective not only transmits the short wavelength UV, but the image in UV is parfocal with the visible light image. For imaging the birefringence of the sperm, we switched the Ultrafluar to a rectified condenser (since the Ultrafluar lenses would not provide anywhere near the extinction factor needed to visualize the weak birefringence of the sperm head). We focused both 'condensers' in turn to project a sharp image of the UV-reflecting micro-mirror superimposed on the specimen (which was illuminated with the 546-nm green light from the primary mercury arc lamp mounted above the Glan-Thompson prism).

In addition, several inches below the micro-mirror and above the compensator and condenser, we mounted a half-inch-thick polished crystal of calcite. This calcite slab splits the light passing through it into two orthogonally polarized beams. Thus, through the ocular, we would see the single birefringent image of the cave-cricket sperm head, superimposed with a double image of the micro-mirror and of the field diaphragm. Only the components above the calcite crystal appeared double because of the double refraction introduced by the polished calcite crystal.

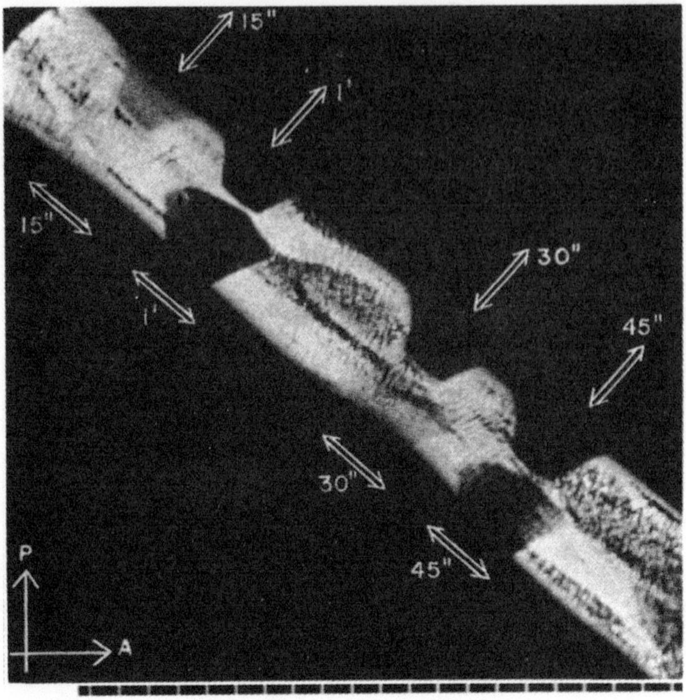

Fig. 3.23 Stretched thread of DNA gel irradiated in pairs by polarized UV of equal intensity, but with opposite E-vector directions as indicated (Inoué S, Sato H, Deoxyribonucleic acid arrangement in living sperm in Hayashi T, Szent-Györgyi A (eds.), Molecular Architecture in Cell Physiology, Prentice Hall, Englewood Cliffs, NJ, pp. 209–248, 1966. [A 24])

An advantage of such an arrangement is that the two images of the micro-mirror have the same brightness and are plane polarized at right angles to each other. So we could irradiate the specimen (initially a thread of DNA) with two small spots of UV (images of the micro-mirror) which have the same incident energy but are polarized at right angles to each other.

What we found right away was most striking (*Collected Works* A-24). In several seconds, the irradiated region of DNA thread (Fig. 3.23) and the sperm head lost its (negative) birefringence and generated a birefringence-free patch, and unlike with the metaphase chromosomes, the irradiated region showed no sign of swelling. With the UV polarized (with its electric vector oriented) parallel to the long axis of the sperm, it took four times longer to introduce the same loss of birefringence as with the UV polarized at right angles to the sperm axis. In other words, UV polarized perpendicular to the sperm axis was four times more effective in introducing the birefringence loss than UV polarized parallel to the sperm axis.

This ratio of 4:1 coincided exactly with the dichroism (absorption ratio for UV of wavelengths from 250 nm to 390 nm) of B-form DNA fibers measured earlier by Seeds and Wilkins in London (Fig. 3.24). When we micro-irradiated stretched gels of pure DNA, the negative birefringence of the irradiated regions again decayed

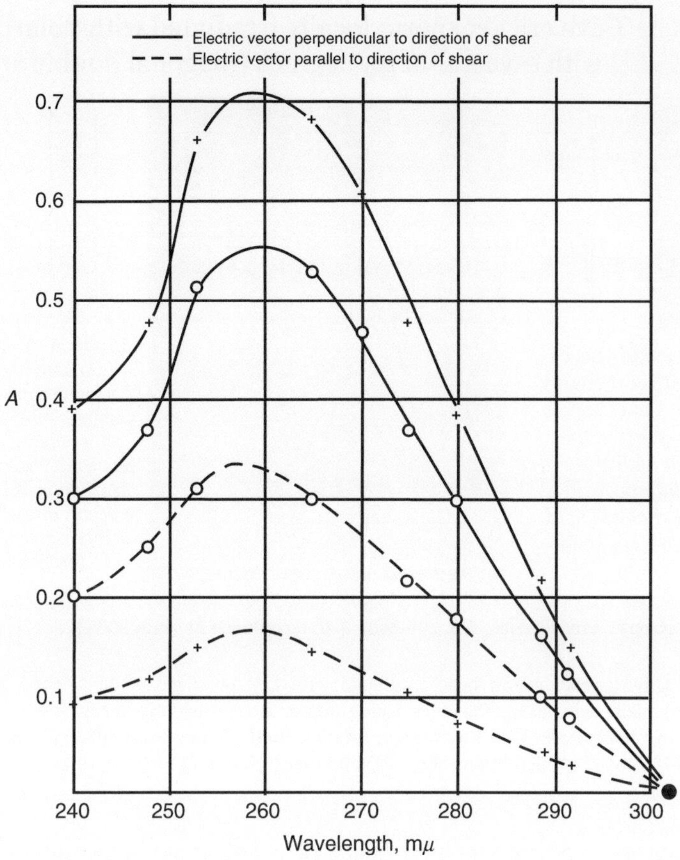

Fig. 3.24 UV dichroism of DNA fiber (absorbance vs. wavelength) determined by Seeds and Wilkins. E-vector perpendicular to shear: +, parallel to shear o (Inoué S, Sato H, Deoxyribonucleic acid arrangement in living sperm in Hayashi T, Szent-Györgyi A (eds.), Molecular Architecture in Cell Physiology, Prentice Hall, Englewood Cliffs, NJ, pp. 209–248, 1966. [A 24])

four times faster when the irradiating UV was polarized perpendicular to the fiber axis than when it was polarized parallel to the fiber axis.

All of these observations could be explained by the arrangement of the nucleotide base pairs which, in the B-form of DNA (the form found in living cells), are stacked like rungs of a (twisted) ladder, i.e., perpendicular to the length of the DNA molecule.

Because the conjugated bonds in the base pairs lie parallel to the plane of the bases, they absorb UV, and refract light much stronger, when the electric field of the light wave is also oriented parallel to the plane of the bases rather than when it is at right angles to it. In other words, UV is better absorbed when its electrical vibration (plane of polarization) is parallel to the base pairs, or at right angles to the length of the sperm head (thus, the 'negative dichroism' in UV). Likewise, light is refracted

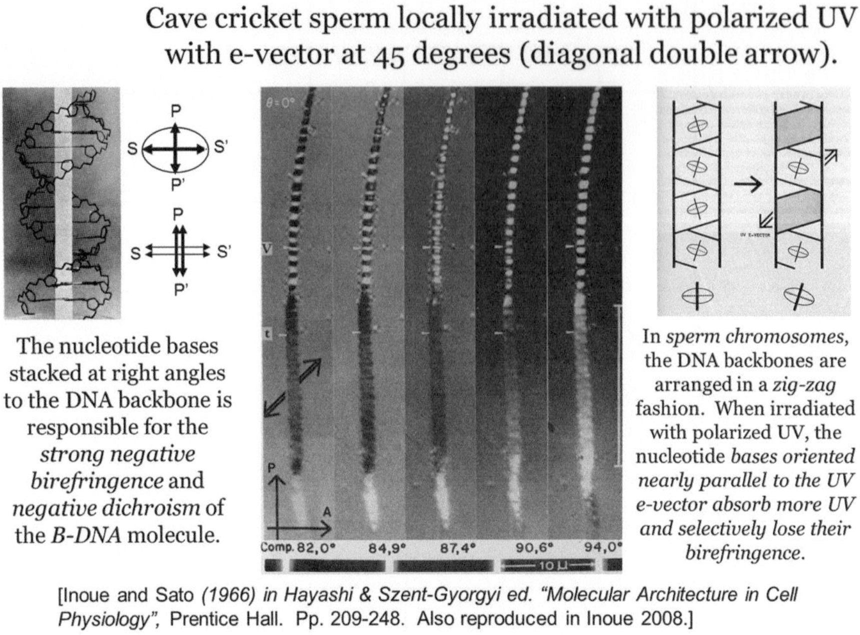

Fig. 3.25 Cave cricket sperm head irradiated with UV polarized at 45° (Inoué S, Sato H, Deoxyribonucleic acid arrangement in living sperm in Hayashi T, Szent-Györgyi A (eds.), Molecular Architecture in Cell Physiology, Prentice Hall, Englewood Cliffs, NJ, pp. 209–248, 1966. [A 24], Inoué S, Video Microscopy. Plenum Press, New York, 1986. [Appendix V as A 40])

more when its electrical vibration (plane of polarization) is parallel to the base pairs, or at right angles to the length of the sperm head (thus, the 'negative birefringence' observed in visible light). Once the bases absorb the intense UV, they change their orientation or composition, and the birefringence observed in visible light decays.

The birefringence of the cave-cricket sperm head (and incidentally of the stretched gel of DNA also), however, showed a zigzag pattern so not all of the DNA molecules in the sperm head (or the gel) were oriented parallel to the long axis of the sperm or of the DNA thread. We were, nevertheless, able to use the same argument to explain the differential loss of birefringence in each zig- and zag-region of the sperm head irradiated with UV polarized *diagonally* to the sperm axis (Fig. 3.25).

To analyze the birefringence changes in those minute zig- and zag-regions, we developed a special method of photographic photometry. The analytical method was based on microdensitometer scans of numerous, uniformly processed 35-mm negatives taken with the compensator and specimen oriented in different directions (Fig. 3.26). It is an analytical method that I am especially proud to have developed, but too involved to describe in detail here. Suffice it to say that I managed to develop a mathematical formulation that checked the validity of the densitometer readings

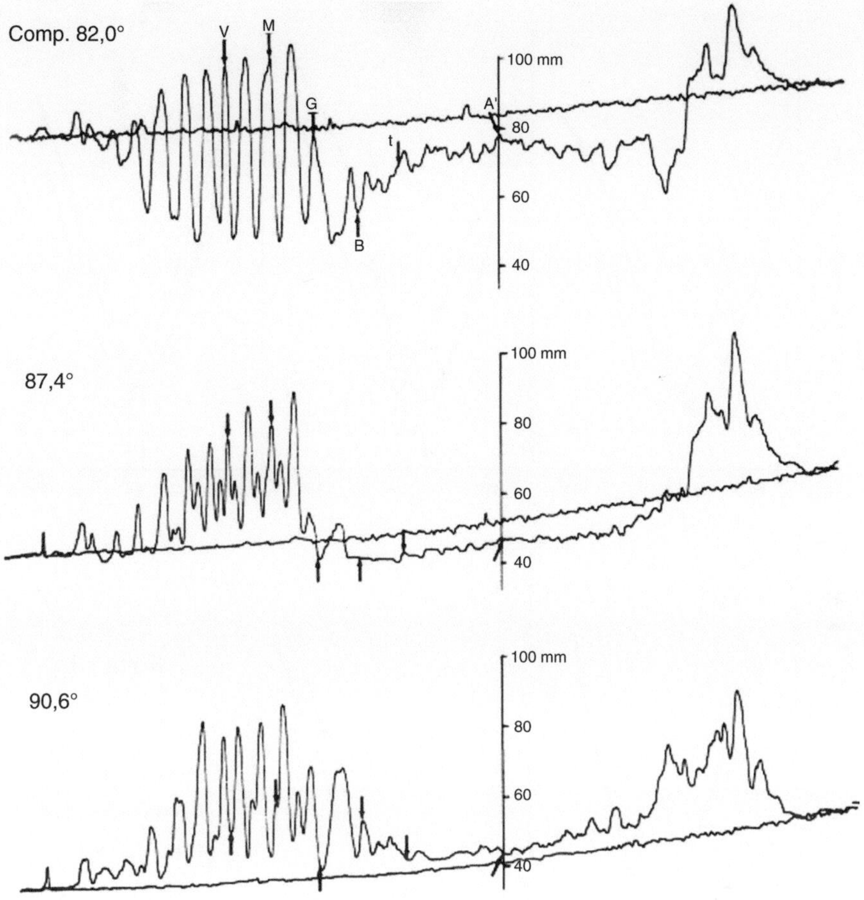

Fig. 3.26 Densitometer traces of locally irradiated sperm head (Inoué S, Sato H, Deoxyribonucleic acid arrangement in living sperm in Hayashi T, Szent-Györgyi A (eds.), Molecular Architecture in Cell Physiology, Prentice Hall, Englewood Cliffs, NJ, pp. 209–248, 1966. [A 24])

at every point along the length of the sperm, while deriving the 'birefringence retardance value' and true 'azimuth orientation' of each irradiated and un-irradiated zig- and zag-region along the length of the sperm head (Figs. 3.27, 3.28, 3.29a, and 3.29b; *Collected Works* A-24).

The outcome was a two-stranded, flattened coil-of-a-coil model for the cave cricket chromatin, with the chromosomes arranged in tandem in a unique sequence in the sperm head (Figs. 3.29b and 3.30).

About the same time as we analyzed the cave-cricket sperm, Herbert Taylor showed, by radioactive labeling of late-replicating chromosomes, that the chromosomes in grasshopper sperm were also arranged in tandem with a unique sequence. On the other hand, as far as I am aware, the two-stranded, coil-of-a-coil

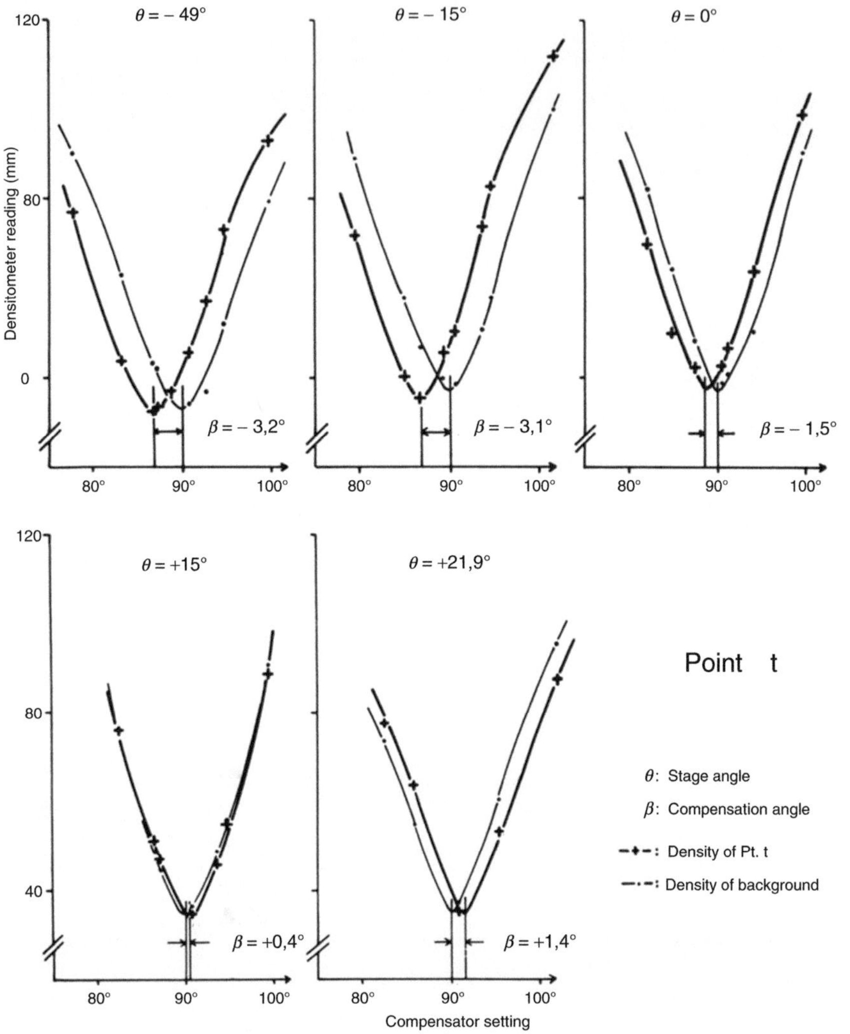

Fig. 3.27 Determination of compensation angle at various stage orientations for point (t) along the sperm, irradiated with UV polarized at 45° (cf. Fig. 3.29a) (Inoué S, Sato H, Deoxyribonucleic acid arrangement in living sperm in Hayashi T, Szent-Györgyi A (eds.), Molecular Architecture in Cell Physiology, Prentice Hall, Englewood Cliffs, NJ, pp. 209–248, 1966. [A 24])

arrangement of the chromatin in sperm has never been imaged by electron microscopy, perhaps such structures fall into an intermediate dimension not readily visible by EM.

The record of these exciting analyses, which we carried out during the last 3 years of our stay at Dartmouth, is published briefly in (Inoué and Sato 1962). It is also described in detail in *Collected Works* A-24, as well as in Japanese in Inoué & Sato, 1966.

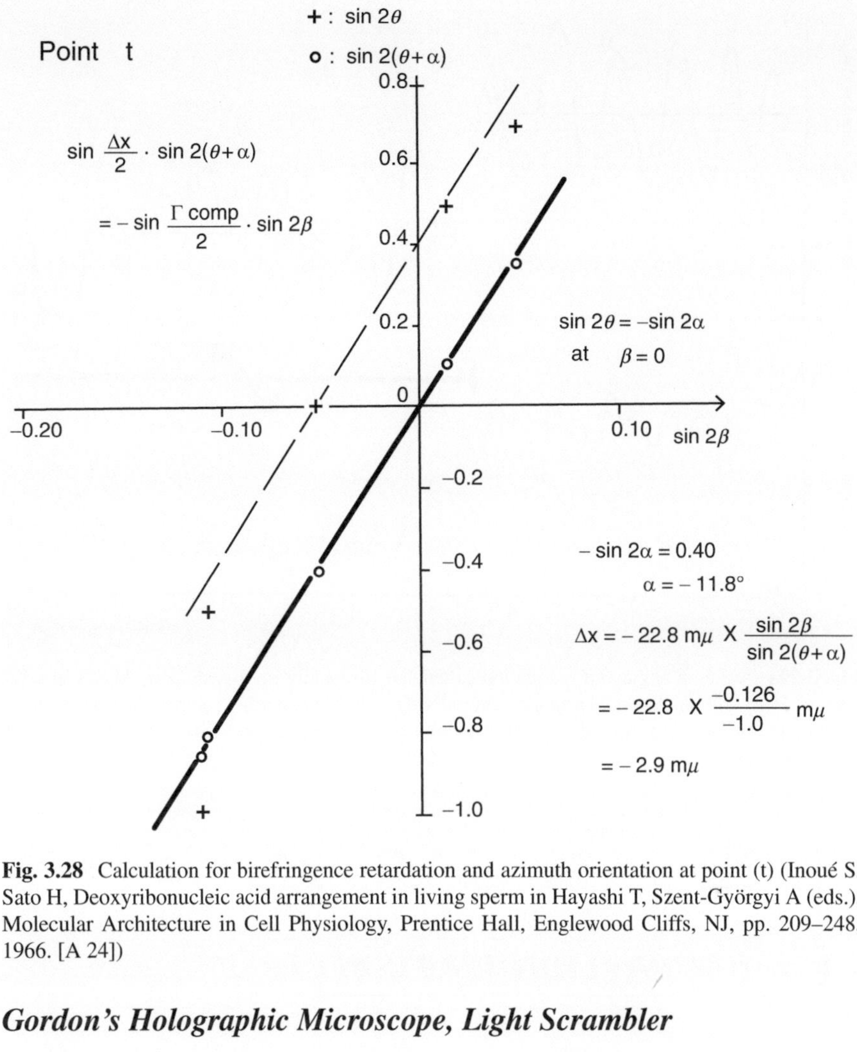

Fig. 3.28 Calculation for birefringence retardation and azimuth orientation at point (t) (Inoué S, Sato H, Deoxyribonucleic acid arrangement in living sperm in Hayashi T, Szent-Györgyi A (eds.), Molecular Architecture in Cell Physiology, Prentice Hall, Englewood Cliffs, NJ, pp. 209–248, 1966. [A 24])

Gordon's Holographic Microscope, Light Scrambler

One day, while we were at Dartmouth, I went with members of our group to a seminar given in the Physics Department concerning some properties of lasers which had just become popular. What was demonstrated by the speaker was the incredibly long coherence length of the monochromatic light wave that was generated by an Argon-ion gas laser. In fact, the coherence length of the emitted beam from the laser was so long that it could be shown to interfere with the wave reflected off the wall of the lecture room some 20 feet away from the laser. The low frequency interference pattern between the two waves (the original wave and the wave that bounced back off the wall, modulated by its vibration) could, in fact, be heard as sound when the two waves were combined and their differences listened to through an amplifier!

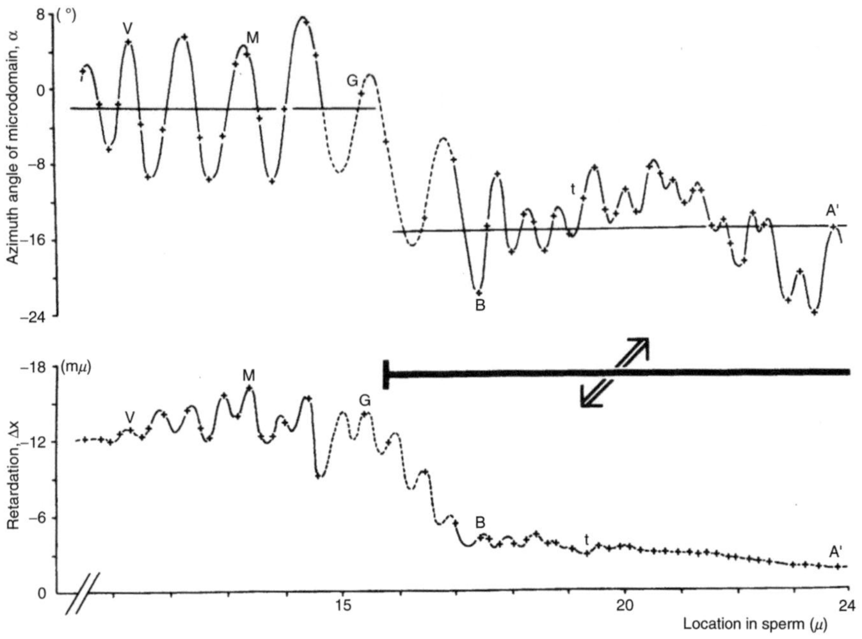

Fig. 3.29a Azimuth angle and retardation calculated for locally irradiated sperm head (Inoué S, Sato H, Deoxyribonucleic acid arrangement in living sperm in Hayashi T, Szent-Györgyi A (eds.), Molecular Architecture in Cell Physiology, Prentice Hall, Englewood Cliffs, NJ, pp. 209–248, 1966. [A 24])

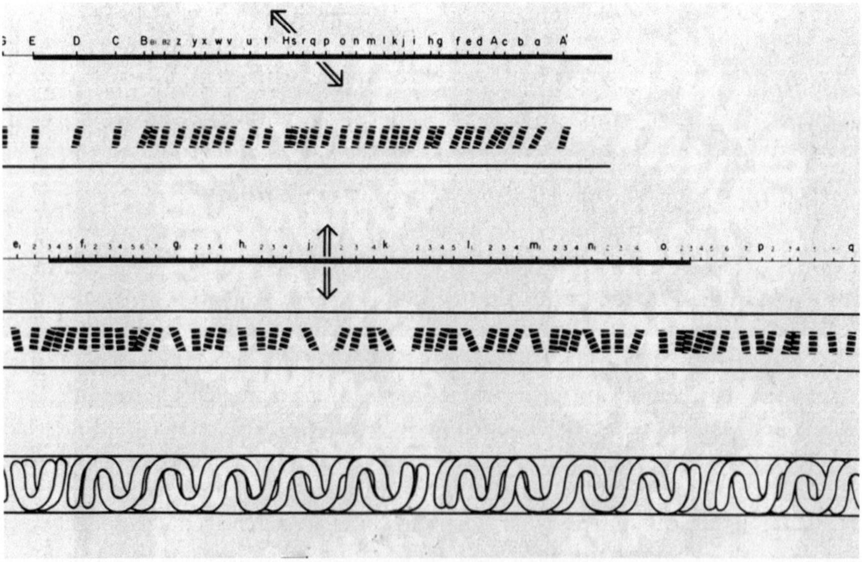

Fig. 3.29b Calculated model (Inoué S, Sato H, Deoxyribonucleic acid arrangement in living sperm in Hayashi T, Szent-Györgyi A (eds.), Molecular Architecture in Cell Physiology, Prentice Hall, Englewood Cliffs, NJ, pp. 209–248, 1966. [A 24])

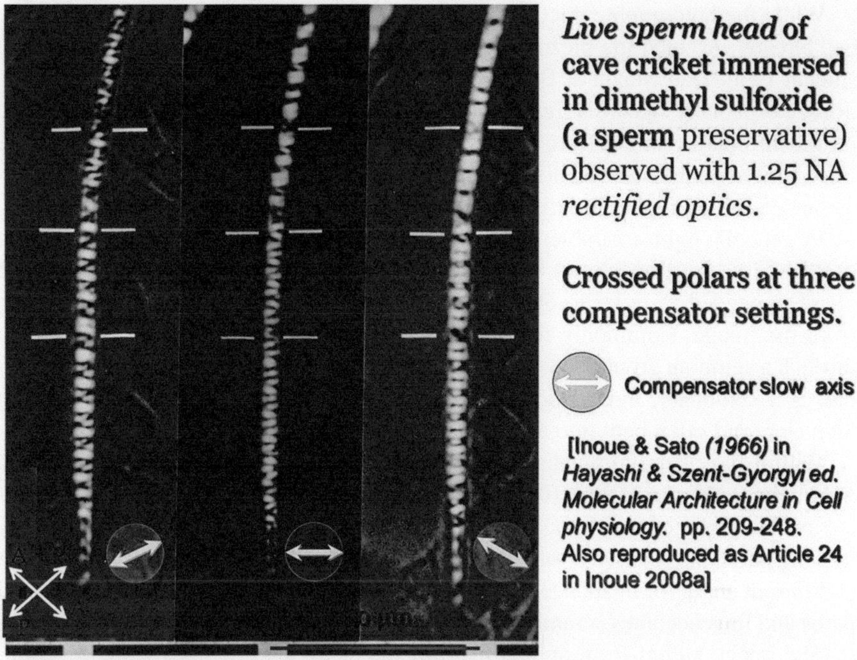

Fig. 3.30 Photomicrographs of un-irradiated live specimen (Inoué S, Sato H, Deoxyribonucleic acid arrangement in living sperm in Hayashi T, Szent-Györgyi A (eds.), Molecular Architecture in Cell Physiology, Prentice Hall, Englewood Cliffs, NJ, pp. 209–248, 1966. [A 24], Inoué S, Video Microscopy. Plenum Press, New York, 1986. [Appendix V as A 40])

Gordon Ellis, who had joined us not too long before then, was struck by the huge coherence length of the laser beam. This led to his considering a new type of microscope whose imaging attributes could be selectively altered *after* an image was initially captured. What he did was to use a laser for the microscope light source and divide its output into two paths. One of the beams passed through the microscope as usual, while the other bypassed around the microscope. The second beam was then directed to be recombined (and aligned parallel) to the first beam after the first beam had exited the microscope optics. Thus, the waves making up the two beams were made to interfere in the image plane and form a holomicrographic image.

After the hologram is captured, the phase relation between the direct and indirect beams is altered and recombined by various means. By selecting appropriate means by which the two beams are recombined, one could generate a dark field, phase contrast, or several other types of final images (Ellis 1966). While exceptionally clever in concept, the final image, or hologram, suffered from optical noise commonly encountered in images reconstructed through holography. Thus, unfortunately, the new system failed to provide microscope users with a new tool that would have been revolutionary.

While the holographic microscope turned out to be disappointing, Gordon, who had already invented the piezo-electric micromanipulator while he was a graduate student (with Dan Mazia at Berkeley), continued to make other ingenious inventions. I especially liked his light-scrambling optical fiber, which makes the intense (monochromatic) output from the laser a useful light source for microscopes. The Ellis light scrambler is exceptionally simple in design, yet its highly effective performance is nothing but intuitively obvious (Ellis 1979).

In general, light scramblers are devices for converting the laser output beam (with highly uniform phase) into a beam with random phase. Randomizing the phase is needed in order to eliminate laser speckles and associated optical noise from the image. Commonly this task is accomplished by passing the laser beam through a spinning ground glass, or wedge, or their equivalent. Instead, Ellis' scrambler is a stationary, light-conducting fiber made out of a **single** fiber of uniform diameter (and often bent into a loop).

Within the fiber illuminated with a focused, converging beam from the laser, light bounces back and forth (by total internal reflection) at different angles depending on the angle of incidence of the entering beam. Thus, without involving any moving parts, the focused laser beam (converging onto the entrance port of the fiber at different angles of incidence) travels within the fiber experiencing varying optical paths and thus becomes scrambled.

The image output from optical instruments using laser sources without scramblers suffers from optical noise seen as mottles or speckles covering the field. To eliminate such noise, we now incorporate the Ellis fiber optic light scrambler for all of our microscopes requiring the intense monochromatic illumination provided by a laser source (Fig. 3.31).

Stockholm International Congress

To explore cell fine structure at our Cytology group at Dartmouth, we looked for a creative biologist skilled in electron microscopy. My personal preference was Bjorn Afzelius from Sweden, a protégé of John Runnstrom of Stockholm. Bjorn had shown a striking '9 + 2' arrangement of microtubules in sperm tail of sea urchins and other organisms. So, on the occasion of the International Biophysics Congress in Stockholm in 1961, Sylvia and I took a side trip to the marine lab at Fiskebaekskil to meet Bjorn and his family.

On the drive to the lab, Bjorn, who had picked us up at the bus stop, suddenly braked his car and hopped out into the woods exclaiming "chanterelle." He came back with a handful of the yellowish mushroom that we later savored at their home. While interested, the Afzeliuses preferred to bring their children up in their homeland, Sweden, so I missed out on recruiting Bjorn. Later, he was to discover the correlation between dysfunction of cilia and flagella and *situs inversus*, a rare anomaly in which the location of the internal organs of a vertebrate organism is reversed so that the heart is on the right side and the liver on the left side of the individual.

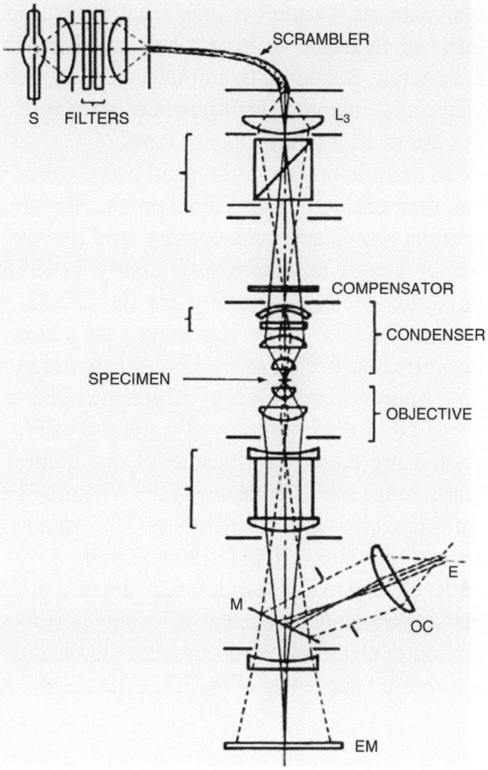

Fig. 3.31 Schematic of Shinya scope-5 incorporating Ellis' light scrambler (Inoué S, Video Microscopy. Plenum Press, New York, 1986. [Appendix V as A 40])

From Stockholm, I went to Germany while Sylvia toured England. I first visited the Leitz works in Wetzlar, where I met W. J. Schmidt from nearby Giesen who joined us for a lunch banquet. Then I went onto the Zeiss works at Oberkochen where I had a quiet dinner with Kurt Michel, and to Wilhelmshaven on the North Sea to visit Roland and Isolde Dietz. In Hanover they had spent a sabbatical year with us and introduced us to crane fly spermatocytes. At Wihelmshaven, Roland was working in Hans Bauer's department at the Max Plank Institute where Haemmerling was also continuing his famous studies on the genetic role of the nucleus, using the large uni-nucleate green alga, *Acetabularia*.

At the Dietz's apartment, Roland was explaining the punctuality of the German TV news broadcast to me. A big clock face on the TV set clicked down tick, tick, tick, second by second, to 7 o'clock. "There," said Roland, but the countdown continued with no sign of the announcer for another minute and a half. Just as Roland exclaimed, "Something terrible must have happened!" the announcer came on with the news that a wall was being erected in Berlin to divide the East and West. That must have been the evening of August 13, 1961, nearly 30 years before the Cold War ended and the Berlin Wall came tumbling down.

Much to the Bauers' and Dietzes' amusement, I tried reaching Sylvia in London by phone, but the hotel operator refused to connect me to her room or to divulge her whereabouts. I hadn't realized that protection of privacy went that far in England! Still we managed to rendezvous at Murdoch and Rosalind Mitchison's place in Edinburgh, Scotland, as initially planned. After touring the Scottish moors where waterfalls emerged from nowhere on top of the grassy hills, we headed home to Plainfield after a brief stay in Paris.

In Plainfield, the children had been looked after by John and Rosemary Marshall and their children from Philadelphia, who enjoyed the country life with the horses, running brook, and fresh country air. I had first met John Marshall, whose work and person I came to admire very greatly, in 1955. He was a fellow finalist at the two-day-long selection interview for the Scholar in Cancer Research of the American Cancer Society at the Harriman estate along the Hudson River. Since then, we became close friends as well as colleagues in science. Together with Woody Hastings and Andrew Szent-Györgyi, John traveled with Hidemi and me to Japan on the occasion of the International Congress of Physiological Sciences in 1965 and even visited my parents and sisters at our home in Ohmori. When we eventually left Dartmouth, John's presence in the Anatomy Department at Penn was a major reason for me to choose that University. Unfortunately, he died in a car accident in Africa shortly after our arrival in Philadelphia.

Returning to our search for an electron microscopist at Dartmouth, we eventually chose Bob Kane, who was a former postdoc with Dan Mazia. Bob was not only skilled in electron microscopy but also had developed newer methods for isolating the mitotic apparatus (Fig. 3.32). It was the major protein in those isolates, made

Fig. 3.32 Bob Kane, Gordon Ellis, and Ray Stephens (from *left to right*) (Courtesy SI)

Fig. 3.33 Shinya with Art Forer, first PhD in Dartmouth Molecular Biology Program

using hexylene glycol, that Ray Stephens analyzed for his thesis. Ray was the first to be recruited to our newly established Molecular Biology Graduate Program at Dartmouth, while Art Forer, who transferred from Rochester with me, was the first to receive a PhD degree since our arrival at Dartmouth (Fig. 3.33).

MBL Physiology Course

In 1963, I met Ray Stephens, who stood out among the many bright students in the Physiology Course at Woods Hole. Before coming to the MBL that summer, Ray, as a biochemistry major graduate student at Pittsburg, had been analyzing the respective roles of hydrophobic and ionic interactions between protein molecules using blocked peptides that he had synthesized.

During the Physiology post-course-project period at MBL, Ray wanted to try an experiment that I had suggested, namely, to use UV microbeam to cut across striated muscle fibrils at different locations to show which connections were necessary for muscle fibers to contract. It was already 8 years since the sliding filament model was published, but this model was too esoteric to be accepted by many experts studying muscle contraction. In fact, Albert Szent-Györgyi even called it the creepy crawly hypothesis. So Ray went around MBL and met with several of the skeptics, including Dick Podolsky, and others who had their own alternate models. Ray had the skeptics agree that if the myofibril would still contract (upon applying Mg-ATP) after severing the protein filaments in this or that portion of the sarcomere, that they would concede the validity of the sliding filament model.

For 2 days after he started the project, Ray was totally uncommunicative. Then he appeared and declared, "The quartz coverslip you gave me is no good." I asked him how he could decide that and he explained: In desperation, he had placed the myofibrils on the outside of the coverslip so that there was no intervening material between the UV focusing condenser and the specimen. Then, as anticipated, the myofibril lost its birefringence, and its optical density, selectively in the region where it was irradiated. Apparently, I had accidentally given him a glass coverslip which wouldn't transmit the 280-nm UV beam needed to sever the protein filaments.

Using a quartz coverslip, Ray then imaged UV coming through a narrow slit to sever the I-filaments on one side of the sarcomere but not the other, and showed that the Z-membrane would be pulled to the edge of the A-band only on the side with intact I-filaments. By permuting the irradiation sites, Ray convinced the skeptics that the only tenable explanation of his irradiation experiments was the sliding filament model (Huxley and Niedergerke 1954; Huxley and Hansen 1954). Those experiments finally convinced the skeptics at MBL that muscle indeed contracted by the mutual sliding of the actin and myosin filaments, rather than by their folding and shortening.

That project, as with the others undertaken during the Physiology Course post-course period, was completed in just 2 weeks! Admiring Ray's performance, I told another of the post-course students that I wished we had bright fellows like Ray for the newly starting Graduate Program at Dartmouth. I was surprised to learn that Ray was contemplating moving, after finishing his block polypeptide work as a Master's thesis at Pittsburg. So I spoke with him and suggested he might expand his muscle microbeam work as part of his PhD thesis with us at Dartmouth. His response: "Yes I'd like to come to Dartmouth, but that project was too easy for a thesis!"

Among the other Physiology post-course students were Bob Barlow, Larry Cohen, and others, who have since become undisputed leaders in their chosen fields. In fact, the MBL Physiology Course that I taught from 1961 through 1966, at Woody Hastings' invitation, was a true eye opener. We had students with all kinds of background from all over the country and the globe (Fig. 3.34). Many were graduate students trained in biology or medicine, but some were also accomplished scientists from other fields contemplating a move into a new field. During the hour-and-a-half lecture first thing in the morning, all of the instructors as well as many from the broad community of Woods Hole filled the Lillie Auditorium, where questions flew left and right from the course students as well as from advanced investigators and experts who came to the lecture.

In one of my early lectures in the Lillie Auditorium, a student questioned an equation I had jotted down on the black board and asked whether the numerator and denominator weren't reversed. The equation was supposed to explain how sound traveled faster along the grain than across the grain of wood, since the elasticity is considerably greater along the grain than across.

Fig. 3.34 MBL 1963 Physiology class

After reading about this property of wood in John Tyndall's 1867 monograph *Sound*, I had devised this audible demonstration to explain the concept of anisotropy using sound, since the concept of optical anisotropy (such as birefringence and dichroism) was not easy to demonstrate directly. I first prepared two blocks of hard wood cut to the same size, one along the grain and the other across the grain. Then, by tapping on the end of the block whose grain ran along its length, I could show that it clicked with a pitch an octave higher than when I tapped on the end of the block whose grain ran across the block. In other words sound travelled twice as fast along the grain of the hard-wood block than it did across the grain (see Section III-3 in *Collected Works* A-40).

My favorite demonstration for anisotropy worked just fine, but indeed as the student had pointed out, I did have the elasticity and density of the material interchanged in the numerator and denominator of the equation that I had jotted down on the black board. The student who asked the question turned out to be Don Glaser, who had already won a Nobel Prize in Physics for his invention of the bubble chamber, which records the charge, mass, and splitting of particles by visualizing their tracks in super-saturated vapor. He was taking the Physiology Course since he was interested in switching over from physics to biology.

For the laboratory sessions which started after the morning lecture and often lasted past midnight, Hidemi Sato, who came with me from Dartmouth, was an indispensable associate. He was loved by the students for his patient, good-humored help as he guided them through practical aspects of microscopy and preparation of live specimen. Some years, Kayo Okazaki also joined us in helping the students in the Physiology Lab. We all had a great time making do with what limited facilities, time, and equipment we had available at the MBL.

In those days before the Loeb Building was built, we taught the Physiology Labs in the north wing of Old Main, a three-story shingled building (Fig. 2.15) located in front of the Lillie Building. Old Main, where Jacques Loeb and others had worked many years ago, was in pretty sad shape by the time we occupied the basement room in front of the marine tanks which were set below ground level. We chose the basement room for proximity to the seawater tanks and for stability of the benches where we lined up the American Optical rectified microscopes for the students.

But one summer, the cracks between the floor boards in the ceiling became so bad that we had to have corrugated plastic awnings built for us above the microscope bench to keep rain water, sand, condensate from the ultracentrifuge refrigeration unit and, after the annual Tarpaulin Cove picnic, salad dressing and stale beer from pouring down through those cracks. Once again, it was Homer Smith who came to the rescue. He had his carpentry crew install awnings overhead to protect the precious polarization microscopes for us.

One weekend in July, towards the end of the Course each year, everyone would board boats and head to Tarpaulin Cove. The lobster and steamed-mussel picnic, the various ball games, and the swimming and strolling on the Cove's beach brought the students, instructors, and their families all together for a memorable day of complete relaxation. During the Course itself, the students occupied the wooden dorms on campus, while the instructors with families were housed in the recently built Devil's Lane cottages. In that way, the Course instructors, Woody and Hanna Hastings, Ken and Barbara Van Holde, Harlyn and Jean Halvorson, Ed and Marion Adelberg, Fred and Sally Karush, and ourselves, together with all of our children, became acquainted as neighbors for the whole summer.

Molecular Biology Program

Returning from our summer experience in Woods Hole to Hanover, we were responsible for teaching the Histology Course to the medical students, and other courses to students in the Molecular Biology Graduate Program. As noted before, Cynthia Knudson and then Bertha Woodward (Fig. 3.35) prepared excellent collections of tissue slides for the medical students under the supervision of Ken Cooper and Hidemi Sato.

We replaced a number of the old formalin-preserved human tissue sections (where much of the cellular features were barely recognizable) with freshly fixed monkey biopsy tissues where the relevant cellular features were well preserved. This emphasis on using physiologically meaningful slides was unfortunately not appreciated by the chairman of the Pathology Department and generated some

Fig. 3.35 Bertha Woodward in her yard with neighbor children

unpleasant friction, although other members of his Department such as Fred and Giesela Dallenbach were all for our approach.

In the meantime, Albert Szent-Györgyi was closing his year-round lab at the MBL. So that gave us the chance to recruit Czuli (Andrew Szent-Györgyi) (Fig. 3.36), who had been working in Albert's lab, to Dartmouth. Czuli would add clear strength to our research efforts as well as to both the medical histology course and the graduate program at Dartmouth. On top of which he was trained as an MD, unlike the rest of us who were all PhDs.

Fortunately, NIH was just announcing a new program for supporting research professorships, which sounded ideally suited for funding Czuli's position. However, when he and I went over the fine print, there were some conditions that would not be acceptable for inviting a scientist of Czuli's stature and interests. So I took the green announcement sheet that came from NIH to Henry Hyle, Associate Dean of our Medical School, and pointed out the problem. To which he commented, "But NIH is about to fund this professorship for you. You shouldn't complain about such conditions." Henry was a friendly member of the Anatomy Department, but neither Czuli nor I were convinced by what Henry said so we wrote our problem to NIH. Very shortly we received a response thanking us for pointing out the problem and stating that the suggested changes would be incorporated into NIH's guidelines. So we were joined, not only by Czuli, but by his wife Eve, and associates Bob

Fig. 3.36 *Right to left*: Fumio Oosawa, Andrew (Czuli) Szent-Györgyi, Ursula Szent-Györgyi, Shinya Inoué

Middlebrook, Forrest Goodall, and Gwen Prior. Eventually Albert's son-in-law Goeff Pollit, who had returned from South Africa, also joined us as Department Business Manager to free me from my distracting administrative chores.

Speaking of university and U. S. government administrators, this example of the green sheet was by no means my only experience in which I felt the NIH and NSF officials were far more open minded and understanding of the researchers' needs than the university administrators. Perhaps as managing recipients of Federal grants, university and research institute administrators feel the need to interpret what they perceive as the regulations literally, but my experience has been that at least some in the granting agencies can be much more understanding of the real needs of the scientists directly involved in research and research training than the managing administrators in the recipient institutions.

I especially recall two outstanding Program Directors, one at NSF and the other at NIH. Bill Consolazio of the Molecular Biology Program at NSF, together with his exceptionally capable associate, Kepi Engel, had a fantastic knack for finding and encouraging promising investigators (to apply for their grants), thereby planting the essential seeds for nurturing the newly developing field of molecular biology. Together with Bill Jenks, John Singer, Dan Tosteson, and Jonathan Wittenberg, I had the pleasure of serving on their panel from 1962 to 1965.

Charlie Miller, who directed the Biophysical Sciences Training Grant Committee at NIH, was another person who understood and very effectively supported the needs for science and research training. By successfully promoting the need for supporting young investigators in the basic medical sciences as an NIH mission, Charlie contributed immensely to the progress of science and medicine in the United States and abroad. Again I got to know him very well, initially by serving on his panel from 1965 to 1970. Later I believe it was Charlie Miller who recommended me to receive one of the prestigious NIH MERIT Awards for two five-year periods.

At Dartmouth, the graduate programs in the Medical School and the College began to flourish. Guiding the Molecular Biology Program were Lafayette Noda who succeeded Manual Morales as chairman of Biochemistry, Clint Fuller who chaired Microbiology, and myself chairing Cytology. Each Department had recruited an exceptionally strong faculty, and the three departments together with some members of the Physiology, Biology, and Chemistry Departments started a major interdepartmental graduate program. In Cytology the graduate students Art Forer, Ray Stephens, and Joe and Jean Sanger, were joined by several postdoctoral and advanced trainees including Don and Ada Olins, Roland Dietz, Bruce Nicklas, and Paul Green who each spent a sabbatical year with us. In fact, in a few short years, the Molecular Biology Program became so highly regarded that we were chosen to nominate winners for a most prestigious international prize for 1967.

In connection with the interdepartmental program, the College raised funds to build a green house on top of the Biology Building. At the dedication ceremony, President Dickey asked me to say a few words on how the green house helps medical research. I mentioned how the endosperm cells from the exotic African Blood Lily were especially suited for studying the events of chromosome segregation during cell division and that mistakes in the orderly segregation of chromosomes lead to genetic and developmental anomalies as well as to serious diseases such as cancer (Fig. 3.37). Hearing this brief explanation, the members of the donor family were all smiles.

Fig. 3.37 Geoff Pollitt, Bill Jackson, Shinya Inoué, and Hidemi Sato inspecting the *Haemanthus* blossoms

One day in the early 1960s, we were visited by Messrs. Ueno and Yamamoto from Nikon who were touring Europe and the U. S. to consider future developments in microscopy. Mr. Ueno headed the research group at Nikon, and Dr. Yamamoto was well known as inventor of an interference microscope bearing the name Françon-Yamamoto. After a few hours' discussion with Gordon Ellis, Hidemi Sato, and me, we were delighted to learn that Nikon would acquire the AO patent and proceed with producing rectified objectives and condensers.

On the other hand, we were unable to convince our visitors from Nikon of the significant advantages that would be accrued by switching to infinity-focused objective lenses. Many years later, the major microscope manufacturers did switch to infinity focus for their biological research microscopes. Ironically, while otherwise introducing superb objective lenses, Nikon was the last of the major microscope manufacturers to introduce infinity-focused biological microscopes and did not do so until 1996.

[2/1/03. As I write this chapter, we received the devastating news that Space Shuttle Columbia disintegrated 39 miles above Texas with seven astronauts on board. It happened at 8:00 AM CST, 16 min before their expected landing at the Kennedy Space Center on Cape Canaveral, Florida.]

Spirit of Occum Pond

As in many schools, the Dean of Dartmouth Medical School held monthly meetings with the full group of Chairmen (the 'Dean's Advisory Board') to decide on matters of general interest to the School. In preparation for one of the meetings in 1965, Clint Fuller, Lafayette Noda, and I met with Marsh Tenney to discuss some basic needs for the Molecular Biology Graduate Program. At the ensuing Advisory Board meeting, Tenney reversed the promise he made to us half an hour ago and completely denied the crucial support needed for the graduate program.

Facing this betrayal by our past Dean, the three of us primarily responsible for the graduate program had no choice but to resign our Chairs on the spot. [In retrospect, that may have been exactly what Marsh Tenney, who once declared that "he was not afraid of anything," had wished. In an article in the summer 2004 issue of *Dartmouth Medicine Magazine*, Constance E. Putnam wrote "A Question of Balance" in which the origin and demise of the Molecular Biology Program at Dartmouth is highlighted. According to the article, Burt Mudge, the Dean at that time, seems to have mistakenly recalled that it was he who had asked for us three chairmen to resign.]

Freed from our chairmanships, we concentrated on our research and teaching until one day a year later Ray Stephens related a rumor that the School was about to dismantle our Department. Andrew Szent-Györgyi, with his MD degree, proposed to take charge of our Department, but the Dean's office rejected the offer. The message that we were no longer wanted at Dartmouth being so clear, some 14 of us in the Biochemistry, Cytology and Microbiology Departments moved to other institutions. Mel Simpson and others moved to the new State University of New York at Stony Brook; Clint Fuller to University of Massachusetts at Amherst; Peter von Hippel to Oregon State University; Andrew Szent-Györgyi to Brandeis University; Kenneth Cooper to University of California at Riverside; and Gordon Ellis, Hidemi Sato, and me to the University of Pennsylvania; and so on.

Although a truly traumatic and sad event, I wrote my father in Tokyo how I was most impressed by the outpouring of support from others across the States; some 40 institutions throughout the country inquired whether we wished to join them. One of father's brothers, a Mitsubishi Bank manager in Tokyo, is said to have expressed his astonishment that in the United States one is not bound to a particular employer, but was free to seek alternate employment.

After retiring to Orcas Island in Puget Sound, Wayne Thornburg used to remark that the Dartmouth Medical School was haunted by the "Spirit of Occum Pond." Marsh Tenney's home abutted that Pond.

Today Dartmouth, which started in 1797 as a school with emphasis on training American Indian youths, has become co-educational with many PhD programs in the College that started concurrently with the Molecular Biology Program. The Medical School itself has grown to a full 4-year School offering MD, MD-PhD, and PhD degree programs.

Be that as it may, 1966 left two indelible shadows in my heart. As noted above, our Department of Cytology, together with our dreams and aspirations for boosting science-based medical education at Dartmouth, was killed by the Medical School Administration. Then, on August 6, we lost my mother, who had been serving as official interpreter for Empress Showa. She had succumbed to pancreatic cancer which had metastasized despite 8 months of intensive treatment. [Father narrates his heart-wrenching experiences of those days in Part 3 of his yet-to-be-published '*Sajo no Ashiato*' ('Footprints in Sand'). Also see *Hideko Inoué Memorial Collection* assembled by father and published in 1967.]

Chapter 4
University of Pennsylvania, More on MBL

Philadelphia

University of Pennsylvania

In 1966, Gordon Ellis, Hidemi Sato, and I left our unfulfilled aspirations at the Dartmouth Medical School and moved to the University of Pennsylvania in Philadelphia. As we established the Program in Biophysical Cytology in the Biology Department there, Bertha and Ralph Woodward, Bob Knudson and his family, and Dick Markley moved to Philadelphia with us. Geoff Pollitt, on the other hand, was hired by Keith Porter to become the first business manager for the Biology Department at Harvard.

At Penn, the Chairman of our Department, Alan Brown, had space in the Leidy Lab generously brought up to excellent modern standards according to our design. It had not been renovated since L.V. Heilbrunn and his students had occupied the space in the 1930s. Having designed the Cytology Department quarters at Dartmouth, and especially with Gordon Ellis, Hidemi Sato, and Bob Knudson joining in the design this time, we were able to achieve an ideal laboratory for the type of work we were interested in.

To enable the development and use of the various types of microscopes and culture and observations of living cells, we were able to incorporate features such as: many microscope cubicles and separate rooms whose lighting and temperature could be independently controlled; carefully cleaned, filtered, and temperature- and humidity-controlled air circulating in each space; noxious fume and heat removed at their source through exhaust vents; floors reinforced to reduce vibration and covered with battleship linoleum to minimize generation of dust; windows facing the town traffic covered with 3-inch-thick granite to shut out air-borne vibration; special 'Ballentine' quartz still to generate extra pure water; an Instant Ocean aquarium room set up for culturing marine organisms; space in the green house for growing *Haemanthus* and *Lilium*; etc.

S. Inoué, *Pathways of a Cell Biologist*, Springer Biographies,
DOI 10.1007/978-981-10-0947-1_4

Ralph Woodward looked after the animals in the marine aquarium room and plants in the green house, while Bertha tended to the Ballentine still and culture systems, while at the same time helping to train the graduate students in basic laboratory manners. In fact, she was a surrogate mother to many, including the instructional staff. Bob in the meantime took charge and renovated the biology machine shop, while Dick set up an efficient photographic dark room and archiving system.

My main undergraduate lecture course was in Advanced Cell Biology. This course was attended not only by seniors and graduate students in the Biology Department, but also was open to students from other departments on the campus. One year, two ladies, who worked as technicians at the Medical School and were taking the course, asked to be excused from the final exam. Upon my urging, they did take the exam and passed with credible performance. I realized it was a forbidding experience for them but hope that their accomplishments helped boost their self-esteem.

Earlier in the course, I met Ted Salmon and Phil O'Dowd from the graduate program in the Biomedical Engineering Department in the Engineering School. They took the course to fulfill their departmental requirement to take a couple of Biology Department courses. They, as well as Dave Begg, Keigi Fujiwara, Kip Sluder, and later Jerry Gordon, Dan Kiehart, and Doug Lutz who had majored in biology, became sufficiently enthused with the subject to plan their PhD work in our Program.

To introduce the graduate students to preparations of living cells for microscopic observation, Gordon Ellis, Hidemi Sato, and I started a laboratory course which we named Biophysical Cytology. John Aronson, who had been a PhD student with me at Rochester and who was now working at the Wistar Institute located across from the Penn campus, came to join our class as did George Langford. George, at the urging of Jean Dan (who was spending a sabbatical year with Tay Hayashi at the Illinois Institute of Technology), had joined our group as a postdoctoral fellow after finishing his PhD thesis in Tay's Department.

Early in the course, in the fall each year, we drove with the class to the New Jersey Coast near Atlantic City. Here the students learned how to net *Dissosteira*, a powerful flying grasshopper that we brought back and cared for in the lab. It surprised me how this outing turned out to be the very first insect netting experience for many of the students who were city born. In the lab, these grasshoppers provided optically clear spermatocytes. As the spermatocytes flowing out from the testicular lobe were spread out onto the bio-clean coverslip into a thin spiral layer, the cells became neatly arranged in developmental sequence of sperm formation (see *Collected Works* A-71). Spread as a single cell layer under halocarbon oils, which are quite permeable to oxygen, they continued to divide for many hours.

For his thesis, Ted Salmon (Fig. 4.1) chose to develop a high hydrostatic pressure chamber through which he could observe the birefringent spindles in dividing cells. This was a highly challenging undertaking, since the windows of the chamber would have to not only withstand the hundreds of atmospheres of hydrostatic pressure, but also not interfere with observation and measurement of the weak birefringence of the spindle in polarized light. Ted, with his engineering background, and

Fig. 4.1 Ted Salmon working on his PhD thesis, using his hydrostatic pressure chamber, with Shinya in MBL L-328 (*Collected Works* A-27, A-58) (Courtesy SI)

with input from Gordon Ellis, did succeed in designing a 3-cm-diameter observation chamber connected to a pressure gauge. The internal pressure of the chamber could be instantly raised to 100 s of atmosphere, or back to one atmosphere, simply by a slight turn of the handle of a stop cock connected to the closed system (Fig. 4.2, lower left). Owing to the low compressibility of water, a tiny shift of volume would dramatically change the internal pressure in the system.

The chamber allowed Ted to make the very first direct observation of chromosome movement accompanying depolymerization of spindle microtubules induced by hydrostatic pressure (Fig. 4.2, upper row). In addition, he succeeded in plotting the birefringence and chromosome velocity versus pressure curve, using the metaphase-arrested *Chaetopterus* spindle (Fig. 4.2, lower right). These measurements expanded the thermodynamic analysis of the microtubule-tubulin equilibrium, which we had earlier developed as a function of temperature alone. Salmon's work introduced the other important state parameter: pressure.

By the time Ted Salmon was making these observations, I had set up a year-round lab at the MBL. In Lillie Room 328, shrouded with a dark curtain across the window, Ted Salmon shared a bench with John Fuseler. John, another student working on his dissertation with me, was measuring the decay and recovery rate of birefringence upon temperature jump, using spindle microtubules in *Chaetopterus* oocytes and fertilized eggs in a star fish, *Asterias*, as well as endosperm cells of a bass wood tree, *Tilia americana*. Unlike the metaphase-arrested *Chaetopterus* spindles, those in the star fish eggs and bass wood endosperm cells continued to divide from metaphase into anaphase, so that in addition to measuring the kinetics of

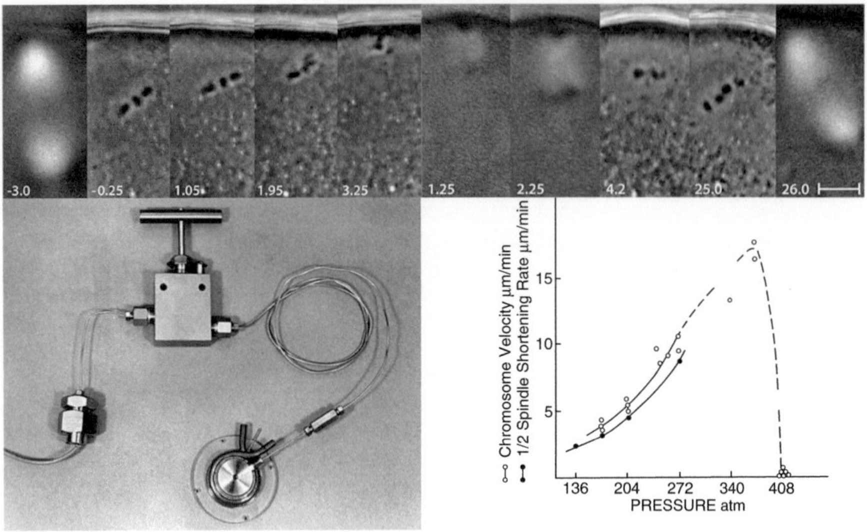

Fig. 4.2 *Top row*: Induced movement of chromosomes in metaphase-arrested *Chaetopterus* oocyte as the spindle depolymerizes (*left 5 frames*), following the application of 3000 psi, and then re-polymerizes (next 5 frames) following release of pressure. Frames indicate time in minutes after pressurization and release. Bar = 10 µm. From: Salmon, 1975. *Lower left*: Salmon's hydrostatic pressure chamber attached to pressure-control valve. From: Salmon & Ellis, 1975. *Lower right*: Velocities of chromosome movement and of microtubule de-polymerization as functions of pressure. From: Salmon, 1976 (Inoué S, Fuseler J, Salmon ED, Ellis GW, Functional organization of mitotic microtubules: Physical chemistry of the in vivo equilibrium system, Biophys J 15: 725–744, 1975. [A 27])

depolymerization and assembly of the metaphase spindle microtubules, John could observe the changing response of the microtubules to cold treatment at different stages of mitosis.

By the time these experiments were progressing in the early 1970s, Bob Knudson had moved to State College in Pennsylvania where he and Cynthia preferred to raise their children outside of the urban environment of West Philadelphia. At State College, Bob constructed houses, while Cynthia made designs for re-modeling. About the same time, Dick Markely became disenchanted with the high crime rate in West Philadelphia and returned to Vermont.

At Bob's recommendation, Ed Horn from the Physics Department Instrument Shop joined our group and oversaw the Biology Shop. In addition to the many, exceptionally high precision components that Ed designed and fabricated for Shinya-scope 5, he designed and made the special constant-flow-rate valve (Fig. 4.3a) for the temperature control slides that we used in Woods Hole.

The ingenious valve allowed us to instantly switch the temperature of the anti-freeze perfused through the slide (Fig. 4.3b), without the slightest change in the flow rate of the coolant (separated from the live cells by a thin coverslip), which would otherwise have interfered with our observation and measurements (Fig. 4.4).

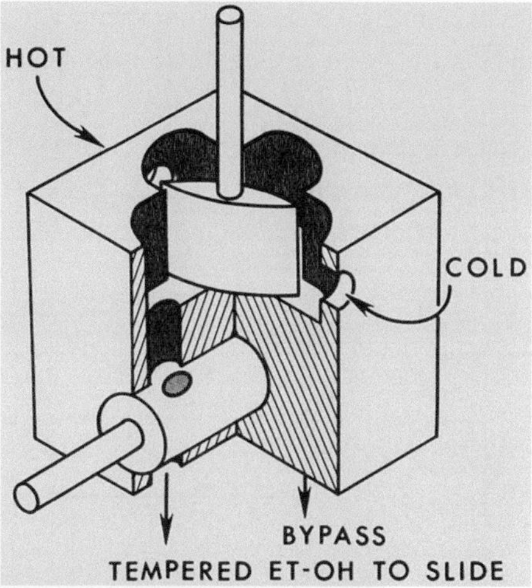

Fig. 4.3a Constant-flow-rate temp control valve (Inoué S, Fuseler J, Salmon ED, Ellis GW, Functional organization of mitotic microtubules: Physical chemistry of the in vivo equilibrium system, Biophys J 15: 725–744, 1975. [A 27])

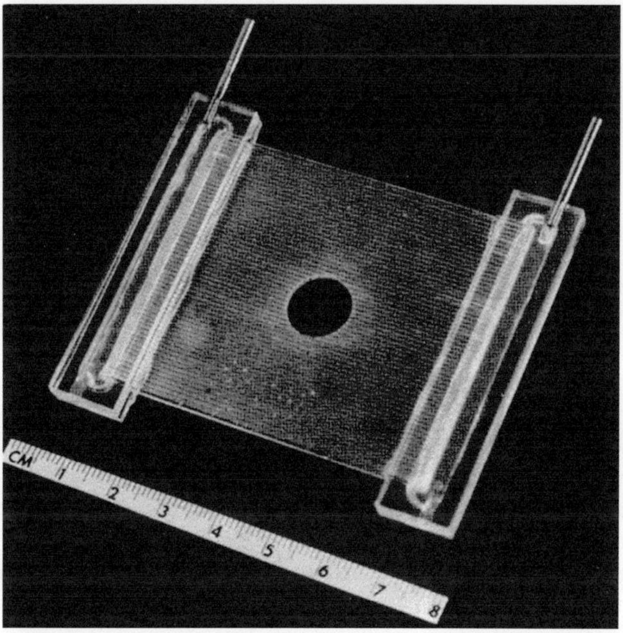

Fig. 4.3b Temp control slide (Inoué S, Fuseler J, Salmon ED, Ellis GW, Functional organization of mitotic microtubules: Physical chemistry of the in vivo equilibrium system, Biophys J 15: 725–744, 1975. [A 27])

Ted Salmon went ahead and showed the kinetics of depolymerization by hydrostatic pressure of purified microtubules isolated in the test tube (in addition to those of spindle in living cells), thus further substantiating the notion that observation of birefringence change in the living cell, induced by change in state parameters, was,

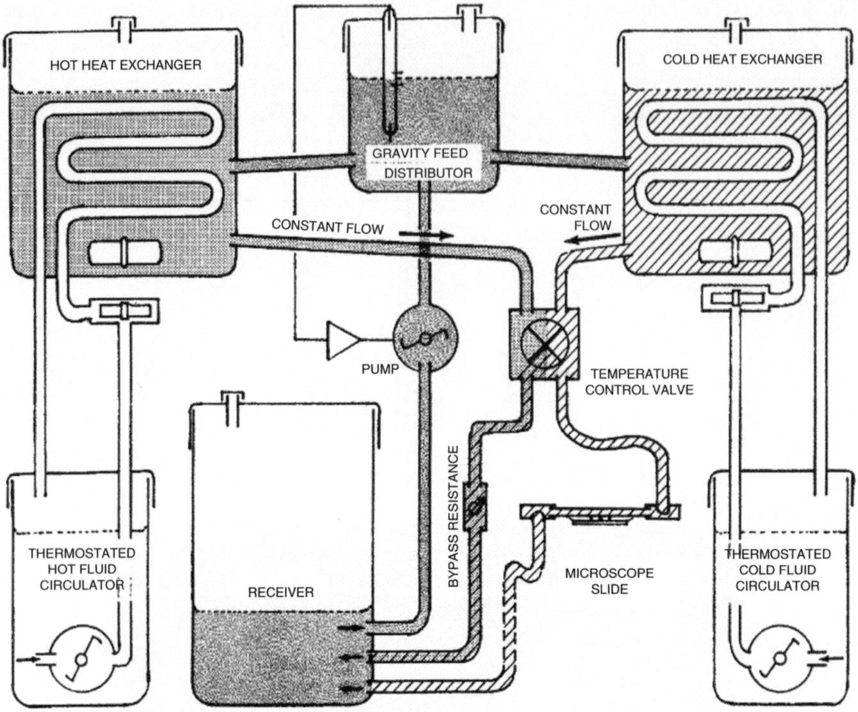

Fig. 4.4 Schematic of constant-flow-rate temperature controlled slide system (From Fig. 5 in *Collected Works* A-27.) (Inoué S, Fuseler J, Salmon ED, Ellis GW, Functional organization of mitotic microtubules: Physical chemistry of the in vivo equilibrium system, Biophys J 15: 725–744, 1975. [A 27])

in fact, directly measuring their effects on assembly and disassembly of microtubules (Salmon 1975).

He further showed how the pressure-induced shortening of spindle fibers, and the accompanying pulling of chromosomes, became faster the greater the hydrostatic pressure. Importantly, however, the fiber shortening and chromosome movement abruptly stopped beyond 400 atmospheres (Fig. 4.2, Lower right) (Salmon 1976). Just as in my earlier colchicine experiment, too rapid a depolymerization of the spindle microtubules simply made the fibers fall apart without performing mechanical work. That observation lent additional credence to the notion that ***slowly depolymerizing microtubules could pull cell organelles such as chromosomes*** (Fig. 4.5). However, as described earlier, it took another 30 years before such a heretic notion was generally accepted.

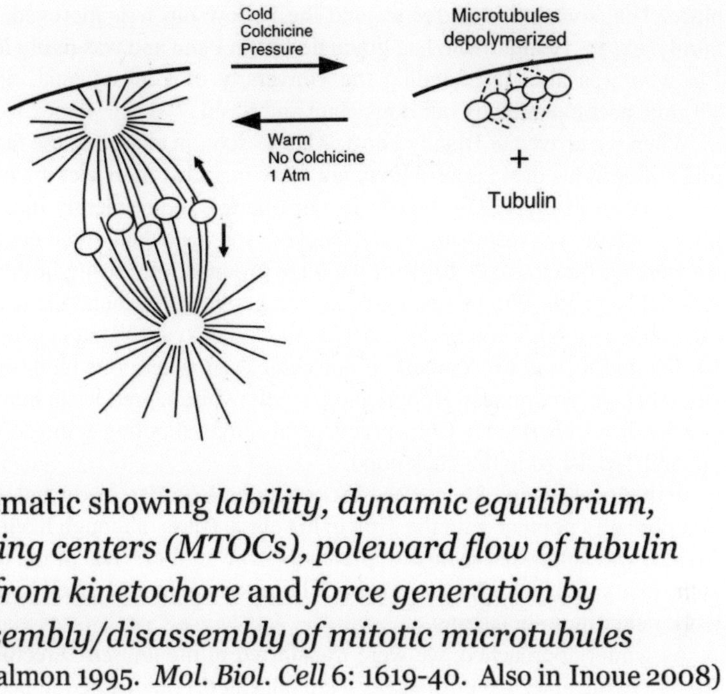

Fig. 4.5 Schematic diagram of labile mitotic spindle (*Collected Works* A-58) (Courtesy SI)

Sabbatical

Before continuing further on our activities at the University of Pennsylvania and MBL, let me digress and recollect the sabbatical year in 1971. By then, we had successfully started our courses at Penn, and the research and training programs were progressing in the freshly renovated Biophysical Cytology labs as well as in Woods Hole. But after the trying efforts at Dartmouth, I was exhausted and even had to be hospitalized briefly for a stomach ulcer. The time seemed ripe for a sabbatical leave, which I had never taken before. At Dartmouth Medical School we were told by Dean Tenney that sabbaticals were not needed "since we could take off and work elsewhere any time we wanted to."

So I applied for a Guggenheim Fellowship and called Lenny Rebhun at the University of Virginia in Charlottesville. To Lenny, whom I had known well at Woods Hole, I asked, "Is there a possibility of coming to the University of Virginia for a sabbatical, but at some facility way removed from the main campus?" In a few days, Lenny called back and said that there is a place called Blandy Farm, some 60 miles north of the campus, which used to be a very productive experimental facility where Herb Taylor and others had carried out cell radiation experiments. Currently there were caretakers for the 712-acre farm, but with no significant research taking place. That sounded ideal for us, and the Fellowship was approved, so the whole family, except Heather who had graduated from Penn and was living in Europe, and Jon who was already attending the University of Pennsylvania, drove down to Virginia accompanied by our dogs Hapi and Sashi.

When we arrived at Blandy Farm, Alan Caspar, in charge of the facility, was sitting stiffly at his desk; poor fellow, not knowing what to expect, he may even have had a tie on that day! He showed us our quarters, conveniently located above the library, which was surprisingly well stocked, and the extent of the huge fields where we were all free to roam. Little of the huge grounds were being farmed then, so the vast fields, ponds with bamboo groves, and gentle hills around the residential areas surrounded by boxwood became our family's 'backyard.' That is where Steve, then 14, learned to ride his Yamaha motor cycle, and where our blond-haired Afghan Hound Hapi (pronounced Hoppi) and friendly white-haired Sashi hunted the woodchucks as an expert team. Chris practiced his target shooting at the defunct radiation pit, and Ted fished in the duck pond.

In the local school, Steve was blossoming both academically and socially, being exceptionally popular with the girls in his class. Chris, although having a few good friends, was somewhat reclusive, preferring to wait for Sylvia or me to pick him up at the Dairy Queen for his car ride back home where he immersed himself in homework rather than socializing.

As winter approached, we were transferred to the unused Director's Cottage at Blandy Farm, since the dorm areas were unheated. Although equipped with a wood stove, the Cottage was not well insulated although fortunately none of us came down with pneumonia. But while the family was buried in six feet of snow, ironically they were receiving the 8-mm movies that I was sending back home from my trip in the South Seas. I was on my Guggenheim trip to study 'group interacting systems'. The goal of the trip was to see if I could find a system, presumably at organismic levels, where one could appropriately study the collective behavior of the group as a function of individual behavior–much as one explains surface tension as a function of the molecules positioned near the interface.

I traveled from the Hawaiian Islands through Tahiti to New Zealand, then Thailand to India, with a side trip to Kashmir and Nepal, and returned via Japan, but never came across the system I was looking for. That is, until I returned to Boston. Coming home from Logan Airport, on Route 28, there was this beautiful cluster made up of perhaps a thousand starlings, the cluster being perfectly spherical unless perturbed, and then 'elastically' deforming as though a soap bubble hit by a breeze. Of course, the individual birds are flying in formation somehow recognizing their role in the whole cluster, which itself sometimes divides into parts or merges from smaller groups to form a larger sphere. The individuals must act with lightning speed to accommodate to the perturbation of the whole. Sardines and other fish behave similarly, but how do we get a handle on the many individuals in the group that are responding to their own inputs? And what are their sensory inputs? My conclusion: I'd better stick to the colligative properties of molecules in cells, where at least I can get a handle on their behavior optically.

After the trip, I decided to take advantage of the green house and highly skilled crew at the Blandy Farm to probe the division of chromosomes in pollen grains. Mr. Ware, an expert horticulturist who was running the green house, helped me set up pots of *Impatience* (touch-me-not) which started to bloom beautifully in just a few weeks. With the primitive polarizing microscope that I set up, I looked for chromosomes in the elongating pollen tube which emerged from the pollen grain growing on an agar layer. But before the nuclei divided, the pollen tubes would all burst at their tips. I was getting nowhere even switching to the cellophane trick that I had learned about in Missouri. In desperation, I called up Alex Faberge who told me how he himself had been unable to repeat the cellophane culture, a point I have already recounted in connection with my Christmas visit to the Stadlers in 1948.

After a healthful, relaxing stay in Virginia for close to a year, thanks to Lenny Rebhun's generous help, it was now time to leave the Farm. For a couple of days before leaving, however, Hapi who had been with us since she was a puppy was nowhere to be seen. Still Bertha had driven down from Philadelphia to help us move, so we had to cut our search short. Finally just before leaving, we found that Hapi was in the Ware cottage, where the bedridden Mrs. Ware had adopted Hapi as her companion. So good-bye Hapi, enjoy the free run on the farm and take good care of Mrs. Ware.

Wynnefield

From Blandy Farm, instead of returning to our home on 52nd Street, just off Wynnefield Avenue in Philadelphia, we headed for a rented cottage in Woods Hole on Sydney Street. That is where young Ted and Steve spent 1972–1973 with Sylvia, who had tired of the dangers in Wynnefield. While commuting alternate weeks to Woods Hole, I kept the house going in Philadelphia, together with Chris, who was attending Central High, and Jon, an undergraduate Physics major at the University of Pennsylvania.

Ted entered the Woods Hole School by Eel Pond, the oldest surviving grade school in Massachusetts, which was still holding regular classes in those days. Steve went to the Falmouth High School, which unfortunately had just shifted to 'open classrooms' with neither the students nor the teachers seeming to know what was going on. Thus, it was only after Steve, who had been a star student in Virginia just a year before, had not shown up for a whole month that the school called Sylvia to ask where he was.

At the MBL, Ted Salmon and John Fuseler were forging ahead with their thesis research, ironically perhaps making even better progress with my frequent absences, being freed from my nit-picking comments. One day, in a totally darkened Lillie Room 328, they were helping me observe the bioluminescence from day-old embryos *of Mnemiopsis* (a comb jelly fish). Into the room walked my 8-year-old son Ted with the comment, "Can't see a thing" as he switches on the light. This, of

Fig. 4.6 Futaba and Fumi visit our Philadelphia house (Courtesy SI)

course, ruined the experiment, but gave Ted Salmon and John Fuseler just the excuse they needed to run off to the Fishery's wharf to enjoy the stripper fishing they had been itching for.

Returning to the 1960s in Philadelphia, our home, a beautiful stone mansion built in 1896 (Fig. 4.6), was located between the Beth David Synagogue and the home of the Fragers, a friendly orthodox Jewish family. Unlike the Fragers, many local residents had fled to the suburbs across City Line Avenue, and our mixed neighborhood was becoming a high crime area. The abandoned house in the big lot in front of our home appeared to be a drop off place for stolen bicycles, and unruly kids would climb the stone wall of the Synagogue and frighten Sylvia tending to her vegetables in the back yard. Steve and Chris had been exposed to the decaying atmosphere at the Mann School, and Ted and his friend Evan were accosted by unruly older girls on their way back from the Gompers School.

After we had lost a Yamaha high-fi set, and finding that every other house in our block had been burglarized at least once that year, we went with a big display chart, which I had put together, to present our case to the district police. Having heard our plea, the police did patrol the streets more frequently in their cars, but not by foot patrol as requested by the Wynnefield Residents Association that I represented.

Still, not all was bleak for the family in Philadelphia. Ted, our youngest son, who later initiated Universal Imaging Corporation, was introduced early to computers at his grade school, Gompers. Gompers was chosen by lottery to be one of the first five schools in the city to receive a handful of computers to be used in the classroom. Later he would frequent the Franklin Institute Museum located not far from the Greenfield High School which he attended in Center City. In addition, he started showing interest in astronomy (Fig. 4.7).

Early in 1967, we hosted my father from Tokyo for a 2 week visit. We had lost my mother in 1966 and father was desperate to find a suitable home for Fumi, his favorite grandchild who had lived with him and my mother for most of her life. He felt that her future would be much more assured if she became fluent in English, and where better to learn than in my family in Philadelphia, if Sylvia and I were willing to take on this responsibility? To this we readily agreed. But was there a school that would find a place for a young teenager with limited English skills? Fortunately, the answer was readily at hand, for Heather, intimidated by the huge public high schools in Philadelphia with thousands of students, had opted to attend Friend's Central School, a private day school near our house for her senior year in high school.

Father met with Dr. Bush, Headmaster at Friend's, who was eager to help us out of our quandary and accepted Fumi as a student starting the following September. This proved to be a happy solution, for Fumi thrived in the nurturing atmosphere at Friend's Central and still maintains friendship with both students and faculty after nearly 50 years. Her English skills played an important role in her job at Sacred Heart University in Tokyo and she has just embarked on a new career as a medical interpreter. Father knew best!

Fig. 4.7 Ted showing view through telescope to his mom

It also became a wonderful opportunity for my father and his grandchildren to become acquainted with each other. In his memoir, he fondly recalls the early morning springtime strolls in Wynnefield, hand in hand with young Christopher who acted as a guide and explained the sights to "Grampi."

Chris made several good friends at Central High, many gathering with Steve to dismantle and refurbish old motor cycles in our spacious garage. Steve has since drag raced professionally until winning the National Championship in 1998 (Fig. 4.8).

Sylvia's bee hives provided bountiful supplies of honey from the bass wood tree towering over the front of our house. In the fenced-in backyard were her vegetable and rose gardens, with a blossoming plum tree by the brick walk where Ted set up his first telescope. A huge magnolia tree full of magnificent blossoms greeted us on the porch facing the side yard every spring, as did the full grown rhododendron bushes. Inside the porch was a glassed-in ping-pong room, connected by French doors to the large entry foyer with a pool table, allowing plenty of room for active kids to play in the winter months. On the second floor, overlooking the magnolia, I had a large study with glassed-in display cases next to our bedroom. Then, we were introduced to the jolly Kuklinski brothers, who used their old world skills to regenerate the intricate ornamental plaster work on our dining room ceiling.

Soon after arriving at Penn in the fall of 1966, I was asked to serve on the Mundheim Commission, a University-wide committee headed by Bob Mundheim

Fig. 4.8 Steve on his touring bike

from the Law School. Its charge was to set up guidelines appropriate for institutional reaction to demonstrations of grievances which were engulfing the campuses, not only in Berkeley and Wisconsin but also worldwide. I recall the disruptions by the blaring loud speakers at Nagoya University, where I visited Fumio Oosawa and his Biophysics Department, and the threats of life to the Medical School Dean at Tokyo University, which prevented my visiting or even talking on the phone with the Dean Junnoske Nakai. I already referred to the threats on Katy's life, then President of Toritsu University, as relayed in his memoir.

Most students at Penn were much less violent or adamant about their grievances. It seemed that some of the faculty members, even some serving on the Mundheim Commission, were more insistent than the students on advocating the use of drastic measures to express one's views or grievances. Thus, it surprised me that I had to even argue the dangers of bodily interfering with access to the hospital operating room and that students who wished to use the library had their rights, too. The ground rules we came up with, including under what circumstances and by whom the city police could be summoned, seemed to work for Penn, but apparently were laughably mild for those in Japanese universities. Dr. Nakamura, Hidemi's advisory professor in Kyoto, wrote Hidemi back that such ground rules as the one we had agreed on at Penn would have cut no mustard at his institution.

A few years later, as a senior member of the Biology Department, I also served on the Dean's Advisory Committee for the liberal arts college, which President Meyerson had characterized as a "hole in the donut" surrounded by well-established professional schools. As I worked on his committee, I came to feel deep respect for our Dean, Vartan Gregorian, who was an astute and energetic Armenian immigrant who effectively plugged up this "hole" as founding Director of the Faculty of Arts and Sciences.

One task with which I was charged by Gregorian was to help the merger of the Graduate Program in Biology with those in the Medical and Veterinary Schools under a common umbrella, a charge that sounded completely natural for our own Biology Department. For reasons I could not comprehend, there was some hesitation by the administration of the Professional Schools to merge, but the idea was wholeheartedly welcomed by their faculty members, and so the inter-school graduate program came into being. Unfortunately for the University of Pennsylvania, after being Provost for 2 years, Gregorian left to head the City Library of New York and later to become President of Brown University in Rhode Island. In both positions, he is known for his most successful leadership. I was saddened to hear a rumor that the professional school deans at Penn actually celebrated the departure of Gregorian.

Dave Goddard, who was Provost and whom I knew also as a biologist, told me early on that Penn was not well suited for all students. However, I did not catch the implication of his warning, so all three of our college-age children, Heather, Jon and Chris, enrolled in Penn on faculty children's scholarships. As I was to learn in the course of time, many of the undergraduates at Penn were actually pre-meds, or otherwise headed for professional schools, and tended to have their heads stuck against the grind stone, a lifestyle not suited for our children, but a lesson that unfortunately I learned much too late.

Biophysical Cytology

In our Biophysical Cytology Program at Penn, Dave Begg, who had switched from studying the ecology of littoral snails, decided to work with Gordon Ellis. Using the piezoelectric micromanipulator designed by Gordon, Dave produced an elegant thesis probing the mechanical individuality of spindle fibers and their strength of attachments to the chromosomes. Even in the presence of the glass micro-needle used on the micromanipulator, the spindle fibers in crane fly spermatocytes could be seen clearly in the rectified polarizing microscope, so that Dave was able to demonstrate how the sex chromosomes were differentially attached to the two spindle poles via unusual fibers. While he produced a superb thesis, I was told that (being of delicate nature) poor Dave had nightmares of crane flies castrating him for having extracted their testes for his studies. Already responsible for raising his own family, and being the only member of his wide family relations that had even proceeded to college, Dave had extra burdens on his shoulders (Fig. 4.9). No doubt Bertha Woodward must have helped him through those trying times, as she did Keigi Fujiwara.

Fig. 4.9 (*Left to right, front*) Kip Sluder, Dave Begg, and Ted Salmon

Fig. 4.10 (*Left to right*) George Langford, Keigi Fujiwara, and Dan Kiehart (Courtesy SI)

Keigi (Fig. 4.10), who had graduated from the International Christian University in Tokyo, was also married. He worked on his thesis with Hidemi and eventually became well established as an expert on the behavior of endothelial cells in blood vessels. Again, with his family problems, Bertha was to help Keigi throughout his graduate student days.

Kip Sluder (Fig. 4.9), who also did his thesis research with Hidemi and a postdoctoral stint with Dan Mazia, has become a major figure in the research on centrosomes, which are located at the poles of the mitotic spindle. Those bodies, which are otherwise attached to the nuclear envelope, split into two before cell division and play essential roles in organizing the mitotic spindle and its microtubules. Until she passed away in 2004, Bertha still used to talk of Kip Sluder, who is now professor at the University of Massachusetts in Worcester and co-director of the MBL AQLM course with David Wolf, fondly as Kipper.

In addition to the graduate students, Hidemi's lab hosted visiting investigators from Japan, such as Tokio Kano from Olympus and Hiroshi Takenaka from Nikon. Some of the graduate students such as Larry Mathews, Phil O'Dowd, and Glen Daniels chose not to tackle a PhD thesis, but instead headed for Medical School.

While Ted Salmon and John Fuseler were finishing their theses in Woods Hole, George Langford joined my lab at Penn as a postdoctoral fellow (Fig. 4.10). As mentioned before, he was recommended by Jean Dan who was visiting Tay Hayashi's Department at Illinois Institute of Technology. Jean recognized George as not only a brilliant promising student, but also felt that his future would become even brighter with exposure to biophysical studies on live cells and dynamic molecular systems of the type ongoing in our lab.

As postdoc, George combined the advantages of the new powers of video microscopy with electron microscopy to show how the arrangement of microtubules became modified where the *Pyrsonympha* axostyle was irradiated with a microbeam of blue light (*Collected Works* A-33). He then joined Bob Allen, who was developing video microscopy in parallel with us, and visualized the motor activities of microtubules fraying off the side of neurons. Later, as Provost at Syracuse

University, George remained to be a leading authority on the interaction of microtubules with actin filaments, two filament systems in living cells that, before George's contribution, had been thought to act as more or less independent motile frameworks.

During my tenure at Penn, we had many other memorable visitors, who impacted on our research, both at our Program at Penn and in Woods Hole. They included Hope Ritter from the University of Georgia, Gary Borisy from Wisconsin, Lou Tilney from Penn, Fumio Oosawa from Nagoya, Kayo Okazaki and Katy Dan from Toritsu, Kent McDonald from Boulder, Andrew and Wishia Bajer from Oregon, and Yuichiro Tanaka who came as the first Dan Fellow from Japan. Other visitors, such as Issei Mabuchi from the University of Tokyo, Yukihisa Hamaguchi from the Tokyo Institute of Technology, and Dick Woodruff from West Chester State, also worked closely with us and our graduate students.

Dan Kiehart (Fig. 4.10), who had been an undergraduate history major at Penn, became expert in microinjection of protein solutions into sea urchin eggs. That was not because of my ability to teach the technique to him, but just because of my encouragement for him to try a technique that should work. (In fact, as I told Dan later, I myself had failed, after a month of trying to set up a working system that did not leak, even after visiting and taking a lesson from the world expert Yukio Hiramoto at Misaki.) So when Issei Mabuchi arrived in 1982 with an antibody that would prevent cleavage but not mitosis that seemed an ideal reagent for Dan to use for probing the impact of cell cleavage on mitosis.

What Dan found by fertilizing sea urchin eggs injected with the antibody were mitosis after mitosis of the nucleus, uninhibited by the lack of cleavage of the cytoplasm. Many tens of birefringent spindles, some white and others dark depending on their orientation relative to the compensator, thus ended up in a single, uncleaved sea urchin egg. With Gordon Ellis's help, Dan also improved on the microinjection technique and has since used the technique to probe the distribution of motor proteins during early development of *Drosophila* embryos.

One unsolved problem with division in sea urchin eggs is: What does inhibition of the fourth cleavage division do to the vegetal pole spindles that ordinarily migrate to the vegetal pole of the egg and generate the micromeres that form the gonads and skeleton-forming mesenchyme cells (Figs. 2.3 and 2.4)? Does inhibition of cleavage also prevent differentiation of the egg cortex that attracts the fourth division spindle to a particular location, namely, the vegetal pole of the egg? Those are tantalizing questions that would be well worth following up.

Gerry Gordon also did his thesis research with me at Penn. He carried out an elegant piece of research, showing a totally unexpected result: Unlike in crane fly spermatocytes, chromosomes moved ***faster*** when a spindle fiber in a grasshopper spermatocyte (which was drawing the chromosome towards the spindle pole) was locally irradiated with a microbeam of UV (Figs. 4.11a and 4.11b). Unfortunately with Gerry, the challenging encouragement (that ordinarily helped other students) backfired and did little to boost his confidence. In fact, he was completely surprised at the degree of enthusiastic interest that his poster elicited at the 1979 Annual Meeting of the American Society for Cell Biology. To this day, Gerry has not pub-

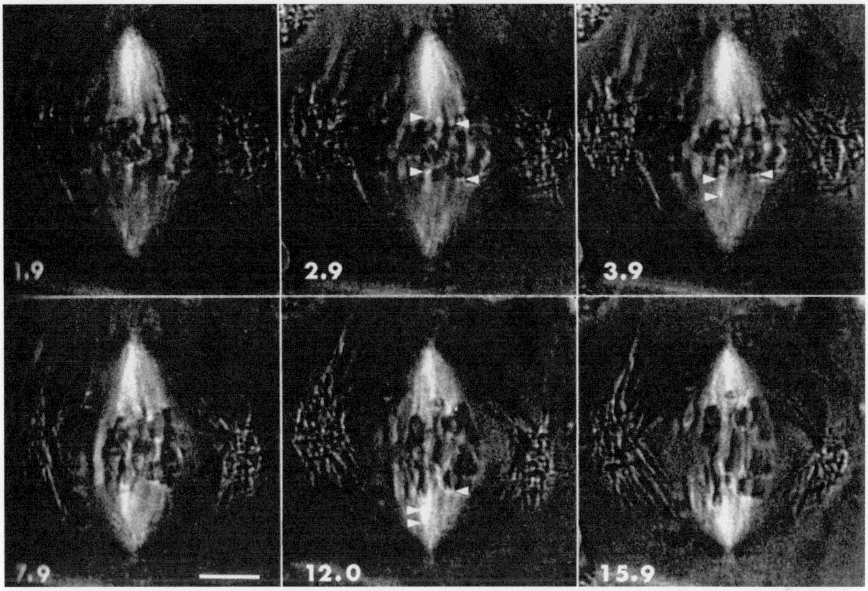

Fig. 4.11a Lower half of Fig. 5 in Gerald Gordon's Ph.D. Thesis (1980): 'The Control of Chromosome Motion: UV Microbeam Irradiation of Kinetochore Fibers' (in primary spermatocytes of a grasshopper *Trimerotropis maritima*) (Courtesy G Gordon, 1980)

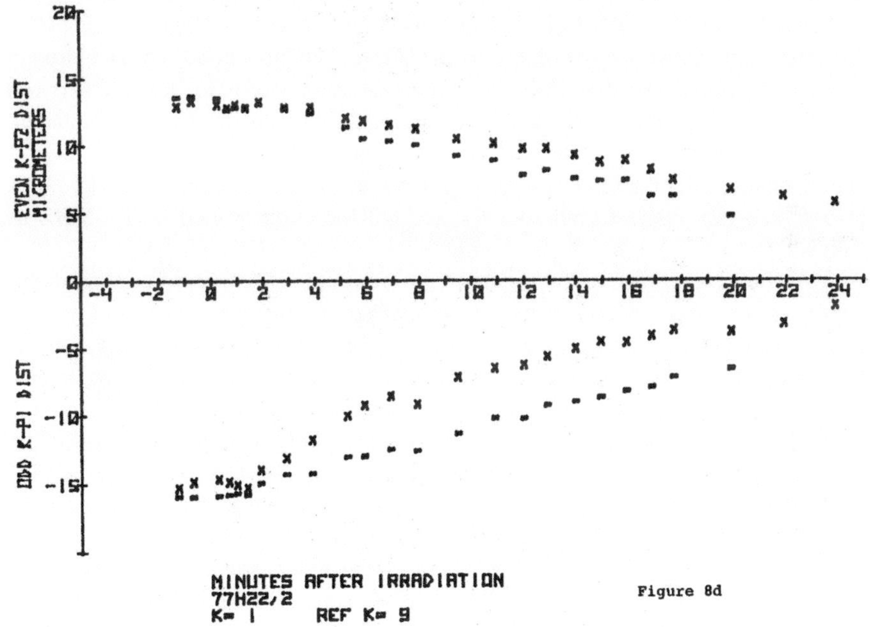

Fig. 4.11b Graph by Gerald Gordon accompanying Fig. 5 above. Transient poleward velocity acceleration by kinetochore whose fiber was irradiated with a UV microbeam (x in lower panel). The fibers connected to other kinetochores were not irradiated (Courtesy G Gordon, 1980)

Fig. 4.12 Doug Lutz and Yuki Hamaguchi (Courtesy SI)

lished his thesis showing his significant finding (*Collected Works* A-32). I am afraid I should not have assumed that all youngsters have the spunk to bounce back when challenged.

Doug Lutz (Fig. 4.12) arrived after receiving a Master's Degree with Erwin Huebner at the University of Manitoba in Canada. Having worked on the cytology of egg formation in an insect *Rhodnius,* Doug was interested in the relation between spindle positioning and generation of cell polarity. Dick Woodruff, a former student of Bill Telfer at Penn, visiting from West Chester State College, and Yukihisa Hamaguchi ('Yuki') (Fig. 4.12), visiting from the Tokyo Institute of Technology, also shared this interest, and so an active collaboration ensued at the MBL as follows.

Yuki, being a protégé of Yukio Hiramoto, was naturally expert in micromanipulation, so that together they managed to displace the metaphase-arrested meiotic spindle in the *Chaetopterus* oocytes from its conventional attachment site at the animal pole of the oocyte. Remarkably, after a brief pause (presumably while new astral microtubules were being formed and searching out the original attachment site), the spindle would swing around and point to, then glide back to, the original attachment site.

More remarkably, when the spindle was turned head over heels with a microneedle, the spindle pole that was initially pointing away from the oocyte's animal pole would now lead the spindle back to the same attachment site. In other words, either spindle pole can find and hook up with the original attachment site, showing not only that the astral microtubules that radiate out from the centrosome are dynamically growing and shortening, but also that they can somehow sense the specific site to which they should attach and bring the centrosome there (*Collected Works* A-45).

As discussed earlier (and as known for nearly a century), the first meiotic spindle is located at the animal pole of the oocyte. As the spindle enters anaphase and then induces cell division at right angles to its length, one ends up with a minute cell–the polar body–which includes the cast-off haploid chromosomes and the cortical attachment site. After the second meiotic division is completed, another polar body is formed while the original one also has divided once more. Thus, one finds three polar bodies at the animal pole of the egg. In contrast, the spindles that are attached to the vegetal pole in the fourth division of sea urchin eggs produce four micromeres that differentiate into gonads and the spicule-forming cells (Fig. 2.4).

The four vegetal pole cells, formed by the third division and which give rise to the micromeres, are quite small, and the division proceeds rapidly in them instead of being arrested in metaphase as in the *Chaetopterus* oocytes. Thus, in order to probe how the spindle was placed at the vegetal pole in sea urchin eggs before forming the micromeres, Doug had to devise a different tact than one that was used to probe the formation of the (metaphase arrested) polar body spindle in *Chaetopterus* oocytes.

For that analysis, he took advantage of an earlier piece of work that John Aronson and I had carried out in 1970. What we had shown was that one could convert colcemid (a synthetic analog of colchicine that is less toxic and more effective than colchicine for depolymerizing the microtubules reversibly) into an inactive 'lumi-colcemid' by irradiation with 366-nm UV light. We had shown that this could be carried out in cells or cell parts that were being treated with micro-molar concentrations of colcemid, so that released from the depolymerizing action of colcemid, tubulin would re-polymerize into microtubules ***locally***. Thus, by using a microbeam to shine a spot of 366-nm UV onto even a portion of a colcemid-treated cell, one could destroy the colcemid in that region selectively and allow the microtubules to grow back in that particular region (*Collected Works* A-26).

After the third division in sea urchin eggs, the nuclei in the vegetal pole cells are normally located near the center of the cells initially but, unless perturbed, soon migrate to the vegetal pole where they become attached and eventually form the micromeres. What Doug found was that those nuclei would remain at the center of the cells so long as the egg was exposed to colcemid. However, if some 10 min after the vegetal pole cells were formed, they were irradiated with 366-nm UV to convert colcemid to its inactive form lumi-colcemid, astral microtubules would form around the centrosome and carry the nucleus to the vegetal pole. Alternatively, if one waited and irradiated the cell a little later when the nuclear envelope was already breaking down, a spindle would form at the center of the cell where the nucleus was located, and then the spindle would migrate to the vegetal pole as its two asters grew and shortened dynamically.

Again, it is the centrosome attached to the nuclear envelope or to the spindle pole that develops the astral microtubules which bring the organelle to the attachment point, whether to the vegetal pole in the fourth division sea urchin eggs or to the animal pole in the oocyte. As the name centrosome given by the cytologists early in the twenty-first Century implies, that body (whose make up and properties are still not well characterized) appears to play a very important, central organizing role in living cells, in part by guiding the formation and activities of microtubules. Today, Kip Sluder, Bob Palazzo, and others are actively probing the makeup and function of this important cell organelle.

MBL Lillie 328

With the Biophysical Cytology lab active at Penn, and the Physiology Course and research projects about to start in full swing at the MBL, our caravan with the family, and with the lab equipment, continued each June from Philadelphia to Woods Hole as it did previously from Hanover, New Hampshire, to the Cape. As usual, the family and pets were crammed into our station wagon, while Dick Markely and Bertha Woodward would drive the rented truck with the delicate equipment, carefully packed and padded. The caravan usually took a rest stop at the "Moo Cow" truck stop at the Rhode Island-Massachusetts State border, where we savored the huge serving of sundaes, and where people and pets alike stretched their legs.

As mentioned before, in the early 1970s, Ted Salmon and John Fuseler carried out their theses research in Lillie Room 328 that I occupied at the MBL. Later, in the 1970s to mid 1980s, Dan Kiehart, Andy Eisen, and Doug Lutz were also to finish their theses research in the same room. Gerry Gordon, who was working with crane fly cells rather than marine eggs, finished his thesis research in Philadelphia.

While John Fuseler and I were working with the reaction of the spindle to sudden drop of temperature, we needed to purge the air around the slide, the condenser, and the objective lenses with dry air in order to keep the moist air from condensing and fogging up the optical surfaces. We managed this by blowing a gentle stream of dry air into a plastic sandwich box, which was perforated on top and bottom and fitted with large plastic washers to accommodate the tips of the condenser and objective lenses. The dry air would fill the box that contained the chilled microscope slide and then escape through the opening around the condenser and objective lenses.

In order to dry the Woods Hole air, which is nearly saturated with moisture especially during the summer, we pumped the air through many turns of an aluminum coil, with a condensate-catching bottle at the bottom, all surrounded by crushed ice mixed with salt. The pump itself was a simple "marine aquarium pump" that gave off the air "puff, puff, puff" from a piston and cylinder driven by an induction motor. The chilled dry air was further passed through a simple desiccator to remove the last trace of moisture before being sent to the plastic sandwich box on the microscope stage.

One day while repairing this system, I asked Bob Tucker, a student assistant from Dartmouth, to replace the tubing that connected the moisture condensing device to the desiccator. When I came back to the lab, he had the system all set up except that the tubing he used was a thin rubber tubing only 3 mm or so in diameter. Surprised that he had chosen such a thin tube, I told Bob, "You're a physics major. Aren't you aware that the flow of air through a tube is proportional to the fourth power of the tube diameter?" Before switching to a plastic tubing with an inside diameter of some 10 mm, we agreed to see how much difference the two tubings would make by collecting the pumped air into an inverted graduated cylinder filled with water.

It's a good thing that we tried the experiment. To my complete surprise, with the same lengths of tubing, the one that Bob had chosen pumped the air several times ***faster*** than the one with three times the bore! So much for my naïve application of Poiseuille's Law. I hadn't taken into account that the Law pertained to steady laminar flow and not to pulsating flow. Nor had I considered the elastic properties of the tubing. With pulsating flow, the resistance to flow must be more like alternating electric current, where one needs to deal with impedances (which is frequency dependent) rather than the type of static resistance encountered in direct current circuits. At least I hope that this explanation is correct. Bob has since gone to medical school and is an active cancer researcher at NIH today.

NATO Summer Schools

While the Cell Biology and Biophysical Cytology courses and research projects were progressing at Philadelphia and in Woods Hole, I was also able to participate in NATO summer schools at Cannes on the Mediterranean in France (1967), Stressa in Northern Italy (1970), and at Szeged in Hungary (1975). On a boat trip out of Cannes is where I first met John Fuseler, who subsequently came and carried out his PhD work with us at Penn. At Stressa, the topic of cell microbeam irradiation continued from Cannes, while the meeting in Szeged related more to cell structure and function with emphasis on cell division and motility. After all, the Medical School at Szeged is where Albert Szent-Györgyi had carried out his famous experiment in 1941/1942 of salt-extracting 'myosin' by grinding up chilled muscle and demonstrating the phenomenon of super-precipitation by adding Mg-ATP (Banga and Szent-Györgyi 1941–1942; also see Szent-Györgyi 1945). Guba and Straub, whose names are also associated with that experiment, had respectively become Dean of the Medical School and Director of the Research Center by the time of our summer school at Szeged.

At the first symposium in Szeged, where Dan Mazia, Phil Siekevitz, and I were seated on the podium as invited instructors, I noted that Dan had removed his tie and looked rather casual. I asked him whether that was on purpose and was told "of course." The reason became obvious over lunch when the younger investigators told us how afraid they were of their own seniors; how they had to watch not getting out of line. Despite some excellent research that was going on (I was especially impressed with the plant tissue culture work), the younger investigators had to toe the line and were not free to express their individuality. This reflection of Hungary, still under the totalitarian communist regime, struck home further when I located what used to be my parents' home in Budapest overlooking the Danube River. The home, which had been taken over by the State, had lost all of its classical elegance, having been converted into crowded tenement apartments housing several families.

Emperor Showa's MBL Visit

In the fall of 1975, Emperor Showa (Emperor Hirohito) came on his first trip to the United States since ascending the throne in 1926. The MBL chose to use the Library catalogue area on the second floor of the Lillie building (now filled with Xerox machines and computer terminals) as the main science demonstration area. So there we set up various microscopes and projection equipment and, of course, live specimen of the local species of hydroids, which were of special interest to the Emperor.

Jim Ebert, who was then President of MBL, had the adjoining reference room converted to a reception area, installed a new desk for signing the guest book, and decorated by fresh flowers and a very fancy cornucopia and tea set prepared by Selma Ebert (Jim Ebert's wife) and her friends. With the scarce funds of the Laboratory, the lavatory area on the second floor of Lillie was spruced up, and the elevator shaft was painted only where the Emperor would see, i.e., just between the first and second floor.

Police sharp-shooters were stationed on Lillie and the adjacent buildings, and the entourage that brought the Emperor to the MBL, and the Empress to the Sandwich Glass Works, were said to have taken the trouble to use un-announced routes, but otherwise security appeared remarkably relaxed. On the sundial park in front of the Lab, Japanese newspaper reporters were correcting the spelling errors on the banners carried by protesters asking the Emperor to protect the whales, and a simple badge handed out to us was all that was needed to enter and exit the buildings.

After having the Emperor sign the guest book in the reception area, Jim Ebert suggested a tea break. Upon which the Emperor slapped his lap and pointed out that he was interested in the science here, not a tea party. So we moved immediately to the demonstration area.

Earlier in the day, Hidemi Sato, who had joined us from Philadelphia for the occasion, busily warned the press photographers not to point their flash lamps towards the mirror on the microscopes that the Emperor would be looking through. Once the Emperor had settled in the demonstration area, Sears Crowell introduced the species of hydroids that had been collected locally and set up for inspection under dissecting microscopes (Fig. 4.13).

On a Nikon rectified microscope that had just won the Industry 500 Award, I had adjusted the half-wave plate to add color to the image of the live hydroid preparation. The image caught the Emperor's fancy, and I received the welcome remark, "How is it that one can see so well through this microscope?" I must have attributed it to the excellent rectified lenses that Nikon had just introduced, but my memory of this third encounter with Emperor Showa is a bit fuzzy. Nevertheless, I still have the dozen or so volumes of the imperial publications that I received in memory of the visit, still wondering which library could make good use of these beautifully illustrated tomes.

Fig. 4.13 Emperor Showa preparing to check hydroid specimen during his visit to MBL in 1975. *Right* to *left* behind his Majesty: Hidemi Sato, Patricia Morse, Marie Abbot, Shinya Inoué (Courtesy MBL)

MBL Prize

Every few years, I had the pleasure of collaborating with Kayo Okazaki (Fig. 4.14), whom I had first gotten to know in 1947 while we were both in Katy's Lab at Misaki. In 1976, she visited us again in Woods Hole, and we followed up on our experiments that we had started earlier: the morphogenesis of skeletal spicules in the pluteus larvae of sea urchins. This won Kayo and me the MBL Prize for that summer.

To recapitulate the sequence of development in sea urchin eggs: in less than an hour after the egg is fertilized and the male and female pronuclei have fused, the mitotic spindle for the first division appears in the middle of the egg (Fig. 2.4). Since the long axis of this spindle is oriented at right angles to the line that connects the animal and vegetal poles of the egg [i.e., the long axis of the spindle lies in the equatorial plane relative to the North and South poles (if we take an analogy from the Earth) and the cleavage furrow is introduced at right angles to the spindle], the first cleavage plane includes the animal-vegetal axis of the egg. The second division spindle is formed in the middle of each of the two equal-sized daughter cells formed by the first division, with their spindles oriented at right angles to the remnant of the first division spindle. Thus, the egg is now cleaved into four, equal-sized cells symmetrically surrounding the animal-vegetal axis.

Fig. 4.14 Katy (Shinya's mentor) and Kayo Okazaki at MBL in 1975 (Inoué S, A Tribute to Katsuma Dan, Biol Bull 187 (2): 125–131, 1994. [A 56])

At the third division, the spindle lies at right angles to both of the spindles of the earlier two divisions, so that one ends up with two tiers of four cells, stacked on top of each other. Of the eight cells thus formed, they may all be of equal size or the four cells facing the animal pole may be slightly larger than those towards the vegetal pole.

Within each of the four vegetal pole cells, the spindle arises parallel to the egg's animal-vegetal axis, and then migrates towards the vegetal pole. The pole-directed ends of the four spindles converge as they approach the vegetal pole, shortly giving rise to four micromeres. Following two successive divisions of the micromeres, they enter the blastocoel to become the mesenchyme cells. The left and right group of the primary mesenchyme cells then each form a syncytial chain (chain of fused cells), within which the seed of the calcareous spicule is formed. As the calcite seed grows its three arms, they become oriented so that within the syncytial chain the growing spicule takes on the form and orientation appropriate for the pluteus of the particular species of the urchin.

As described briefly earlier, Kayo was able to isolate and place into tissue culture a single micromere cell and have it grow into an assembly of mesenchyme cells which would generate the spicule reflecting the species- specific shape. The MBL Prize paper, which follows Kayo's earlier work, is reproduced as Article 29 in the *Collected Works of Shinya Inoué* (2008a). In it, we were able to use a decorating method (by growing micro-crystals of calcite on the isolated spicules whose surface served as nuclei) to show that the axes of all of the micro-crystals were indeed exactly aligned parallel to each other. Thus, we were able to prove that (despite the fact the spicules would not fracture as would a calcite crystal) each spicule was indeed a biocrystal as suggested by W. J. Schmidt.

Hope Ritter: Barbulanympha

In 1978, Hope Ritter (Fig. 4.15), who had previously worked as a postdoctoral fellow with L. R. Cleveland at Harvard, arrived from the University of Georgia, bringing with him wood-eating cockroaches, *Cryptocercus punctulatus*. I was very excited since Hope had spent the past 10 years developing the medium and culture method needed to study the ***living*** nymphidae (the symbiotic protists that reside in these cockroaches).

Hope Ritter first purged the massive, unique glove box (an air-tight container whose contents are reached through a pair of air-tight gloves) that he had brought in a rented car from Athens, Georgia. While purging the space in the glove box, he also purged his culture medium with nitrogen gas scrubbed through a solution of vanadium pentoxide. Purging with the scrubbed gas to completely remove contaminating oxygen was essential, since the nymphidae from the *Cryptocercus* hindgut were strictly obligatory anaerobes and would die in the presence of even trace amounts of oxygen.

In the atmosphere of this oxygen-free glove box, we dissected out the hindgut of the wood-eating cockroach into the oxygen-free culture medium. We squeezed out the nymphs from the hindgut onto a strain-free coverslip and sealed the preparation with silicone grease onto a 'bio-clean' microscope slide.

Fig. 4.15 Hope Ritter (Courtesy SI)

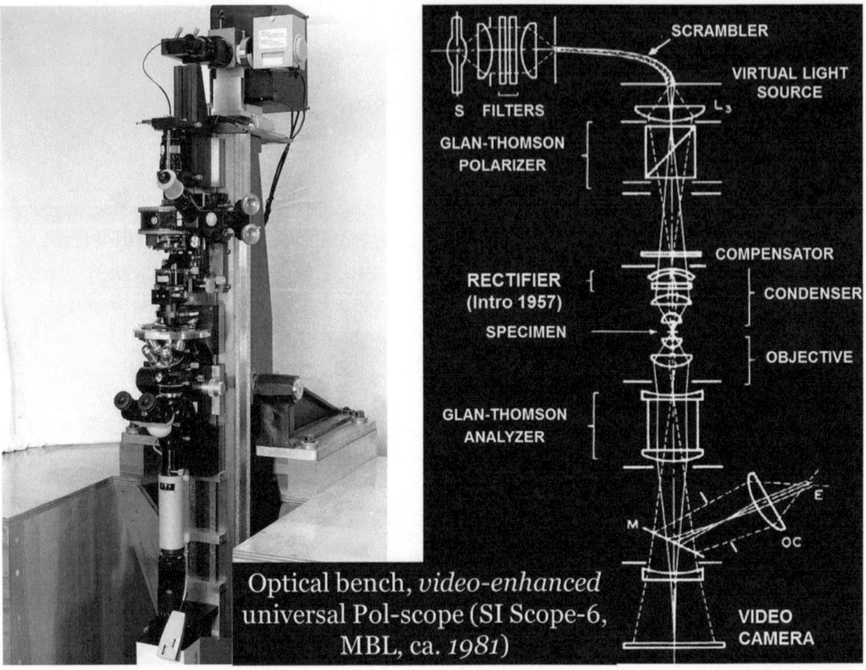

Fig. 4.16 Shinya-scope 6 (Inoué S, Video Microscopy. Plenum Press, New York, 1986. [Appendix V as A 40])

Observed through Shinya-scope 6 (Fig. 4.16), the sight of these nymphs was absolutely spectacular--*Trichonympha* with its massive, flowing flagella ramming and displacing other protists, *Pyrsonympha* with its axostyle whipping through the medium, and the giant *Barbulanympha* cruising slowly, all amidst the background of smaller protists including the spiraling spirochaetes.

A few years later, in 1979, these were the photogenic players in one of the first video tapes that we took using the video microscope that we had just developed. During Hope's first visit, we had not yet realized the usefulness of video image enhancement, so we recorded onto several rolls of 35-mm film the sequence of events taking place during mitosis in live *Barbulanympha*, alternately in polarization and DIC microscopy. Fortunately the records were focused just at the right plane, so that on the photographic records we were able to capture even more than we had noticed by watching the events under the microscope. Thus, they provided striking new information in addition to excellent documentation of the unusual events that we observed.

In *Barbulanympha,* a huge, highly birefringent spindle developed just above the nucleus, between the two "elongate centrioles" which hung below the rostrum at the front end of the body (Fig. 4.17). As our collaborator Donna Kubai was to show with her elegant electron micrographs, the giant spindle was made up of several thousand, well-aligned microtubules (Fig. 4.18).

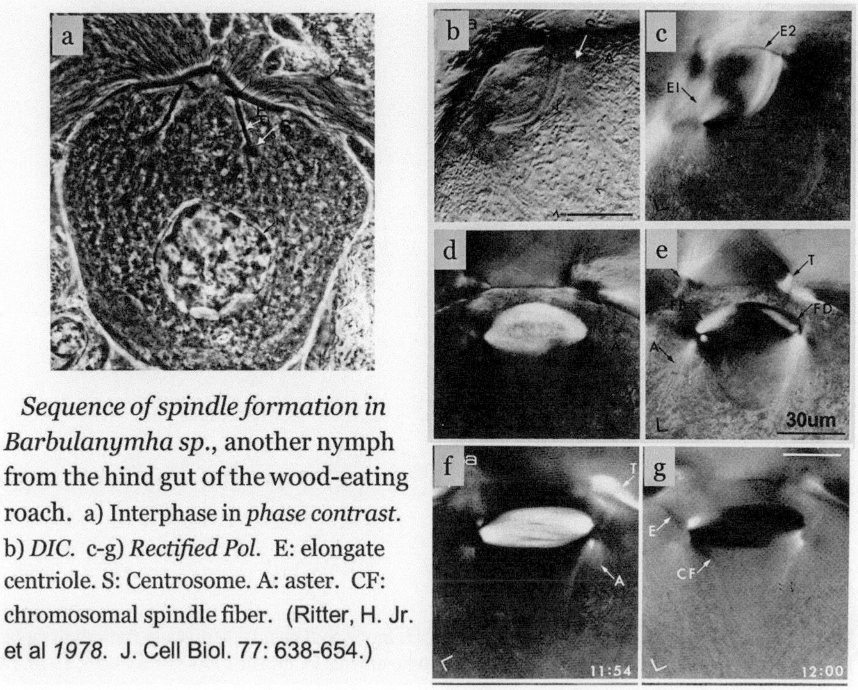

Fig. 4.17 Spindle formation in *Barbulanympha* (Ritter H, Inoué S, Kubai D, Mitosis in Barbulanympha. I. Spindle structure, formation and kinetochore engagement, J Cell Biol 77 (3): 638–654, 1978. [A 30])

In addition, a number of tenuous microtubule bundles ran from the poles of the spindle to the chromosome kinetochores.

As Cleveland had shown earlier, the kinetochores remained embedded in the nuclear envelope throughout the division cycle of *Barbulanympha*, and wherever a kinetochore was located, the nuclear envelope either pointed out towards the spindle pole or puckered inwards, presumably reflecting traction by the chromosomal spindle fibers or by their pushing.

As in many protists, the nuclear envelope in *Barbulanympha* never breaks down during mitosis. [See Bělăr, K. (1926) for illustrated classifications of the source of mitotic organelles in a large group of protists—his figures from this article are reproduced in Fig. 3 in Comment 1 in Supplemental Material to Inoué (2008b).] Instead of being broken down, the nucleus is gradually hoisted up towards the front end of the organism, until the nucleus is finally wrapped around the spindle as though a hot-dog bun around a sausage. At the end of this process, the top of the nuclear envelope is sewn together to form a new seam (Fig. 4.19).

But then how are the chromosomes going to be separated, and how are they going to be packaged into two daughter nuclei? Remarkably, a small opening appears at the top of the nucleus in the middle of the recently formed seam, as though parting of lips (Fig. 4.20, middle and right). Then, the lips open wider and wider, the opening progressing down towards the bottom of the nucleus but with the

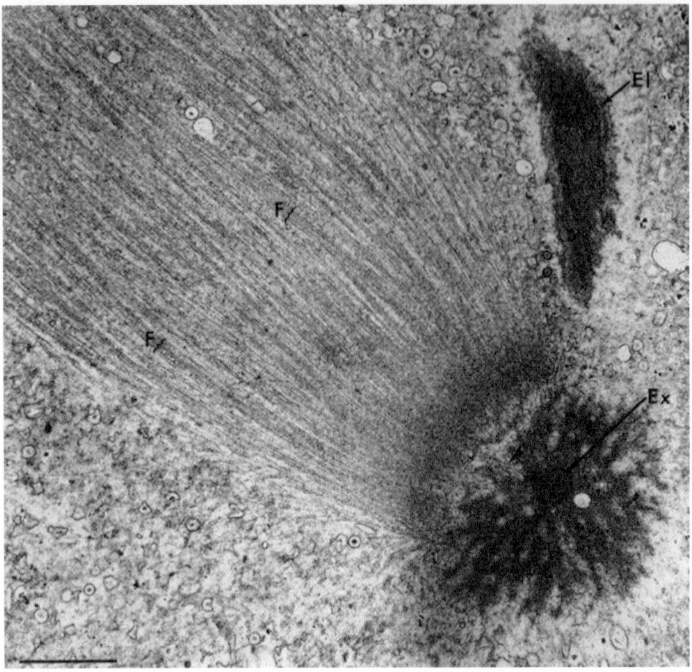

Fig. 4.18 EM of *Barbulanympha* spindle (Ritter H, Inoué S, Kubai D, Mitosis in Barbulanympha. I. Spindle structure, formation and kinetochore engagement, J Cell Biol 77 (3): 638–654, 1978. [A 30])

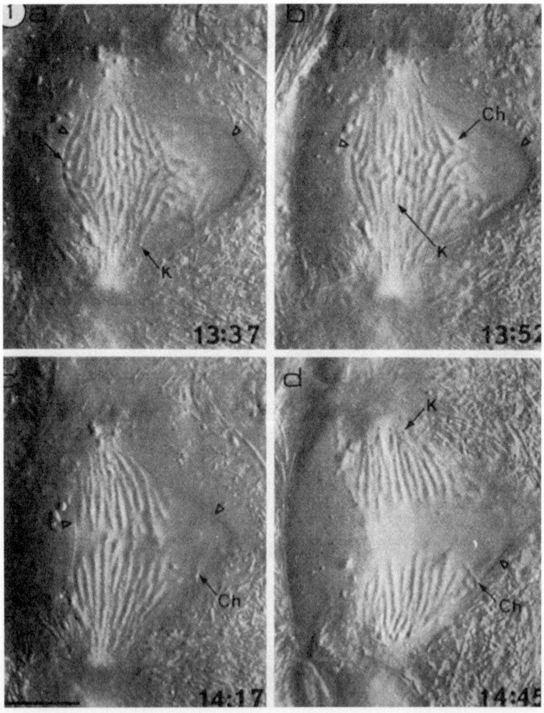

Fig. 4.19 Anaphase chromosomes in *Barbulanympha* (*Collected Works* A-31) (Inoué S, Ritter H, Mitosis in Barbulanympha. II. Dynamics of a two-stage anaphase, nuclear morphogenesis, and cytokinesis, J Cell Biol 77 (3): 655–684, 1978. [A 31])

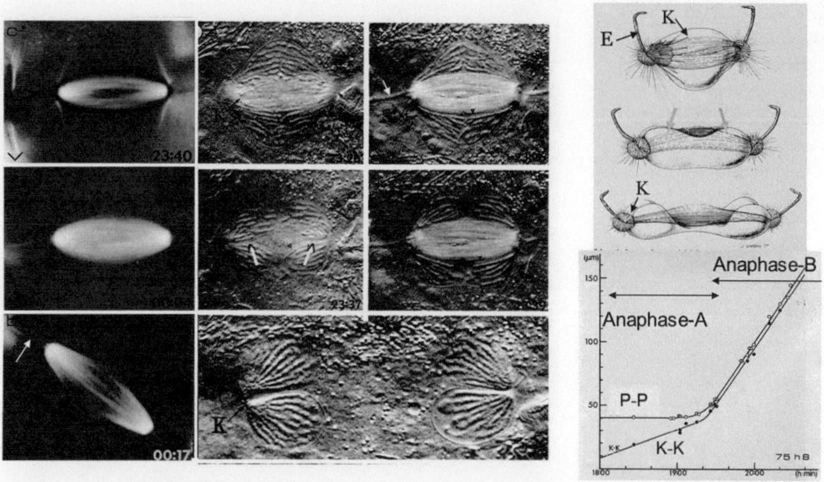

Barbulanympha in Anaphase A & B. ***Kinetochores (K) are embedded in*** ***persistent nuclear envelope which now sandwiches the spindle.*** Sequential *Pol* (left) and *DIC* images (middle). Focus on upper surface shows the *parting seam in nuclear envelope* (between arrow pair in middle column). Top right: schematic drawing showing "nuclear morphogenesis". Bottom right: measured 2-stage anaphase.

Figs. 4.20 and 4.21 *(Left)* Pol and DIC images of live cell (Inoué S, Ritter H, Mitosis in Barbulanympha. II. Dynamics of a two-stage anaphase, nuclear morphogenesis, and cytokinesis, J Cell Biol 77 (3): 655–684, 1978. [A 31]). *(Right)* Schematic of *Barbulanympha* in Anaphase A and B (Inoué S, Ritter H, Mitosis in Barbulanympha. II. Dynamics of a two-stage anaphase, nuclear morphogenesis, and cytokinesis, J Cell Biol 77 (3): 655–684, 1978. [A 31])

seam still holding together near the spindle poles. In the meantime, the chromosomes are pulled to the spindle poles by the shortening microtubules, so that eventually their kinetochores form a beautiful necklace surrounding the narrowed spindle at the poleward surface of the nuclear envelope (Fig. 4.20, middle-bottom; Fig. 4.21; *Collected Works* A-31). Eventually, the tenuous tube of the nuclear envelope (that became longer but persisted in the lower part of the nucleus) finally pulls apart as it seals itself, so that now there are two daughter nuclei still pierced by the spindle. As the spindle continues to stretch to several times its original length, the daughter nuclei are pushed further and further apart, with the chromosomes still surrounding the ever tenuous spindle remnant at its poles.

We called such a crazy behavior of the nuclear envelope, nuclear morphogenesis, and we noted in addition that the movement of chromosome towards the spindle poles was separated from their subsequent movement which was induced by the extreme elongation of the spindle (by as much as five fold)! These were but two of the stories that *Barbulanympha* had to tell. Years earlier, in 1939, Hans Ris had, in fact, shown a case in an insect spermatocyte where chromosome-to-pole movement took place ahead of spindle elongation, but the extreme case of *Barbulanympha*

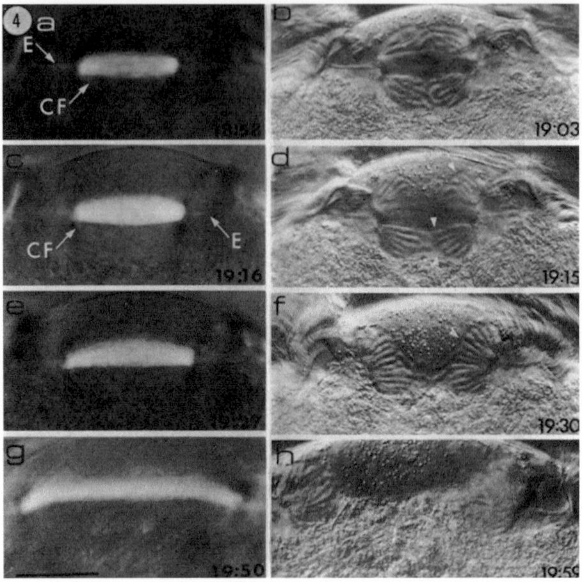

Fig. 4.22 Metaphase through late-anaphase spindle and chromosomes in *Barbulanympha* (Inoué S, Ritter H, Mitosis in Barbulanympha. II. Dynamics of a two-stage anaphase, nuclear morphogenesis, and cytokinesis, J Cell Biol 77 (3): 655–684, 1978. [A 31])

mitosis led us to name the movement of the chromosomes to the spindle pole, Anaphase-A, and their further separation by the increase of the spindle pole-to-pole distance, Anaphase-B (Fig. 4.22). This terminology has since become widely used for description of anaphase events in general.

In two papers in the Journal of Cell Biology in 1978, we documented the phenomenon of nuclear morphogenesis by photomicrographs in live *Barbulanympha* (*Collected Works* A-30, 31). But to this day the exact mechanism that drives nuclear morphogenesis is still unclear. Could the shortening chromosomal spindle fibers, tugging on the kinetochores embedded in the nuclear envelope, alone bring about the surprising change in the nuclear envelope? If so, what does that tell us about the mechanical property of the nuclear envelope itself? How can the envelope be so yielding and yet hang on to the embedded kinetochores?

In addition to raising these questions, the nymphidae must have so much more to tell and enlighten us about aspects of living cells that otherwise escape our attention. I hope the nymphs, together with L. R. Cleveland and Hope Ritter's valiant efforts, will soon come to receive the much wider attention that they deserve.

Full Time at MBL

AQLM Course

Having established a year-round lab at the MBL by the beginning of the 1970s, and having finished my 3-year term as Chairman of the Biology Graduate Group at Penn in 1979, I spent more and more time at the MBL. Then, in 1978, Mort Maser, who was Associate Director in charge of research and education at the MBL, suggested that we start a short course on light microscopy. Although he himself was an electron microscopist trained by Keith Porter, Mort thought the time could be ripe to more fully explore the powers of light microscopy, which after all were uniquely suited for directly probing live cells, as well as hopefully the molecular dynamics inside living cells. However, to gauge the interest in such an offering, Mort would first ask Bob Allen to organize a course, and if that looked promising, he would have me announce our offering in Analytical and Quantitative Light Microscopy (AQLM). With Bob's course on Optical Microscopy offered in December 1978 having attracted the requisite number of applicants, Mort and I agreed to announce the first AQLM Course to be offered in late April 1980.

As course faculty, I had Gordon Ellis from Pennsylvania, Ferenc Harosi from MBL, Ted Salmon from North Carolina, Lance Taylor from Harvard, and Bob Zeh from SUNY at Albany join me. Representing the industry were Mel Brenner from Nikon, Dick Chaisson from Olympus, Dave Cohen from Venus Scientific, Stuart Levy from Leica, Ernst Keller from Zeiss, Tony Silvestri from Sony and Crimson Camera, and Richard Taylor from Colorado Video, among others. By the second offering in 1981, the commercial faculty also included Windham Hanaway from Colorado Video, Jan Hinsch from Leitz, Hitoshi Iida from Hamamatsu Systems, Katsuji Rikukawa from Nikon, and Jerry Kleifgen and Paul Thomas from DAGE-MTI. Most of the 'students' were faculty members or postdocs at universities or research institutes, while some came from industry (Fig. 4.23).

In the earlier years, for more than half of the course, we emphasized the basic optics encountered in microscopy and electronic imaging before discussing various new approaches and research results used for analyzing cell fine structure and dynamics. The week-long course ran from lecture and demonstration sessions in the morning to well past midnight as all the participants, students, and instructors alike, became engrossed in the laboratory exercises and explorations.

Fig. 4.23 MBL AQLM Course first class, April 27-May 3, 1980 (Courtesy MBL)

What quickly emerged from both Bob Allen's Optical Microscopy Course and our AQLM Course was that the video equipment, which were initially used in the courses to simply display the microscope field or back aperture images to a large group of viewers, turned out to have capabilities far beyond anyone's dream. By using a video camera (and later a CCD camera) attached to the microscope, together with an image processing computer, we were able to improve images by (a) enhancing contrast, (b) subtracting away unwanted background (including stationary image features to bring out only the changes), (c) enhancing higher or lower spatial frequency image features selectively, and (d) filtering out noise to bring out image features that had been totally hidden from view so we could detect images otherwise far too faint to be seen (*Collected Works* A-34; Allen et al. 1981).

In other words, the video and CCD cameras had capabilities far better than even the manufacturers had recognized, so that electronic image enhancement could bring out the world under the microscope that had totally escaped detection in the past, including the faint diffraction images of single large molecules which were far below the resolution limit of the microscope. Thus, from the two MBL short courses, the new field of Video Microscopy was born (Fig. 4.24).

Ted and UIC

As the MBL microscopy courses were starting, video recorders were also improving so that dynamic scenes could be captured with little trouble. At the courses, digital image processing was initially provided by the stand-alone boxes brought by

Fig. 4.24 Video microscopy course in Mendoza, Argentina, organized by Mario Burgos (*front row third from left*). Ted Salmon (next to Mario) and Rudolf Oldenbourg (*top of picture*) also joined as instructors for this international course (Courtesy M Burgos)

Colorado Video and Hamamatsu Photonics, but my youngest son Ted from Falmouth High School, who liked to play with the computers attached to those boxes, quickly recognized the potential of a more versatile computer-controlled system.

As Ted entered Cornell University, we purchased an image processing computer board developed by Bob Wang at Imaging Technology in Woburn, MA, following advice from Gordon Ellis. Using the Imaging Technology board, Ted assembled his first image processing computer. Soon following the encouragement from Heather's husband Dick Russell, and with help from Dave Cohen of Venus Scientific, whom I got to know well through the AQLM course, we set up a family cooperation, Universal Imaging Corporation (UIC), using a home equity loan on our house at 40 Shore Street in Falmouth. In 1984, since Ted was still in college, I was the nominal President of UIC.

Half way through Engineering School at Cornell, Ted started to organize a computer group for UIC involving two of his classmates, Jerry Rubinow and Dan Green. Just before graduating, he also recruited Dyon Anniballi and Steve Boyd to help further with the imaging program and Marci Platt to join him in taking care of administrative and other business matters for the company. Thus, the Cornell 6 (Fig. 4.25) became involved in this new venture.

In 1987, Ted took over the Presidency of UIC, but their early engineering site was still at Ted's apartment at Cornell or his bedroom at our Shore Street home in Falmouth and then, for a while, in rented crowded space at 600 North Jackson Street in Media, PA. Early financing continued through the home equity loan. However, Dave Cohen was able to negotiate bulk purchase prices for imaging boards from Matrox and arrange a joint marketing venture with Mike Goldberg of Analytical Imaging Concepts. So from 1988 to 1991, the UIC business, selling both the Image-1 program and computers optimized for the program use, started to become profitable. Bank loans were being paid off and employees finally were being paid more than minimum wage!

Fig. 4.25 The Cornell 6. *Left* to *right*: Ted Inoué, Steve Boyd, Jerry Rubinow, Dyon Anniballi, Dan Green (plus Marci Platt) (Courtesy SI)

In 1990, UIC moved to what appeared to be a huge space in Brandywine Industrial Complex in Westchester, PA, right next to Commodore Computers. However, as they started growing, they had to take over an adjacent office and again double the space in the late 1990s. That space was to be tripled by the time they moved to their current Downingtown, PA, location, and the number of employees had grown to 50 by the time UIC was sold to Molecular Dynamics in 2002.

UIC now distributes their MetaMorph series image-processor/analyzer/system-controller worldwide and even runs a school for educating their employees and users who are new to the systems. In addition, Jeff Stuckey, past CEO and Manager of UIC, introduced the "High Throughput Screening System," a computerized, fully automatic microscope system essential for drug screening. This system measures, tabulates, and screens the reaction of living cells (placed in microwells) to drugs or other agents being tested. Of the Cornell 6, all but one were officers or in charge of R&D programs when UIC was sold to Molecular Dynamics.

The success of UIC obviously reflects Ted's leadership capacity and his and his colleague's technological brilliance. In addition, the company's success in this highly competitive field was also due to Ted and the staff's collegiality with the employees and customers. Ted himself emphasizes that their success was due to the unique collaboration between the UIC technical people and demanding users of microscopes such as me, Ted Salmon, and others. My son continues to believe that "Designing software that was actually used for real research allowed us to optimize the system so that it was highly efficient for lab work. If it wasn't for this feedback, I'm sure Image-1 would have been nowhere as good as it became."

I still smile as I recall the numerous occasions in the earlier days when I asked Ted to introduce a particular image processing or analysis function and his response was, "That's impossible; the imaging board isn't made to do things like that." Then, after a week of silence, he'd come back with a way around the roadblock and show me a single keystroke that achieved what I wanted. This must have happened well over a dozen times while Image-1, predecessor to the current MetaMorph program, was being developed; a real credit to Ted's engineering ingenuity and a joke on me who simply hated black boxes and computers.

Some of the individuals graduating from our MBL microscopy courses started using the image enhancing capabilities of Image-1 for their ground-breaking research. By tracking the movement of individual microtubules on a lawn of putative motor protein molecules, Ron Vale and others discovered a series of kinesins, new motor proteins that propelled microtubules in directions opposite to dynein, earlier discovered by Ian Gibbons in Hawaii using dark filed microscopy. Many of us also used these devices for recording other dynamic cellular events which had been difficult or impossible to visualize in the past. Thus, the new field of Video Microscopy, which started as a product of the MBL microscopy courses, quickly started to bear fruit.

Sperm Acrosome

As prelude to successful fertilization, a single sperm cell must successfully carry its nucleus into the egg cell. And before a sperm can enter an egg, the sperm and egg must unite by first establishing proper contact of their cell membranes. How this contact is established between the swimming sperm and the egg, which is surrounded by jelly and other extracellular coats, was discovered by Jean Dan at Misaki in the early 1950s (see Chap. 2, 'Jean's Return to Science').

Using the phase contrast microscope, that she had brought back home from the States, and a borrowed electron microscope, Jean found that a thin protuberance, the acrosomal process, appeared at the tip of star fish sperm in a few seconds when the sperm was exposed to seawater containing egg jelly. Furthermore, with the phase contrast microscope, she saw the protruding acrosomal process penetrate the egg jelly, reach the egg cell surface, and establish the needed link between egg and sperm. Shortly thereafter, Laura Colwin, who had been Jean's classmate at Penn, and her husband Arthur, who had visited Jean in Misaki together with Laura in 1953, used the electron microscope to show how the sperm membrane at the tip of the acrosomal process, in fact, united with the egg cell membrane.

As remarked earlier, the acrosomal process is so thin that Jean had to use the phase contrast microscope to visualize its presence in live preparations. (For electron microscopy, the preparation is fixed chemically, and therefore the specimen is no longer alive). But by the early 1980s we knew that by using video microscopy we could explore dynamic events in unfixed, living material and in objects which were much thinner than the limit of resolution of the light microscope.

One of the early projects using video microscopy, that Lew Tilney and I investigated, was to record exactly how the acrosomal process elongated and to explore the molecular mechanism that gave rise to the elongation of the acrosomal process. Lew Tilney, a colleague of mine at the University of Pennsylvania, had already made elegant studies of the structures of the sperm and its acrosomal process in *Thyone briareus* using thin section electron microscopy.

Lew had chosen the sperm of *Thyone*, a local sea cucumber found in Woods Hole, because of the very long acrosomal process that could be formed, triggered either by introducing egg seawater (seawater which has been exposed to the egg and which, therefore, contains some of the jelly substance secreted by the egg, which triggers formation of the acrosomal process), or by increasing the calcium ion concentration in seawater. His electron micrographs showed a bundle of actin filaments (the same actin that is found in the I-band of striated muscle) inside the extended acrosomal process, with its growth end pointing to the tip of the process.

Lew's micrographs also showed that the acrosome, which gives rise to the acrosomal process, is a two part structure lying at the front end of the sperm located in a pocket in the nucleus (Fig. 1 in *Collected Works* A-37). Occupying the front portion of the acrosome is the acrosomal vesicle, a small membrane-bound spherical body, whose forward membrane contacts the inner tip of the sperm cell membrane. And surrounding this vesicle is a cup-shaped, periacrosomal region of the acrosome, which contains the unpolymerized actin.

When the sperm is activated, the actin molecules become polymerized, starting at the anterior tip of a tiny 'actomere' which is located behind the acrosomal vessicle. From there, a thin but stout actin filament bundle pushes the acrosomal vesicle membrane from its rear, thus inverting the vesicle membrane. The growing bundle of actin filaments continues to push into the acrosomal vesicle, drawing in the inverted rear membrane of the vesicle, so that the rear membrane eventually protrudes from the front of the acrosomal vesicle to become the membrane that envelopes the elongating acrosomal process. Thus, the thin growing rod of actin pushes the vesicle from its rear and everts the vesicle inside out. Then, over the next few seconds, the acrosomal process continues to elongate for tens of micrometers in length until it reaches its target, the egg (if the egg is, in fact, present).

Viewed in video contrast-enhanced DIC (we used Smith-type differential interference contrast microscopy), the extending acrosomal process of *Thyone* was spectacular (Fig. 4.26). Once the high-calcium seawater reached the sperm (which fortunately stuck to the microscope coverslip only by its midpiece so that its head and tail were free to swivel around), the sperm tail started quivering, the acrosome 'popped,' and the slender acrosomal process grew and grew over the next several seconds. In fact, the process initially elongated at an incredible speed of nearly 10 μm/s, until it gradually slowed down as it became some 50–75-μm long in 7–8 s.

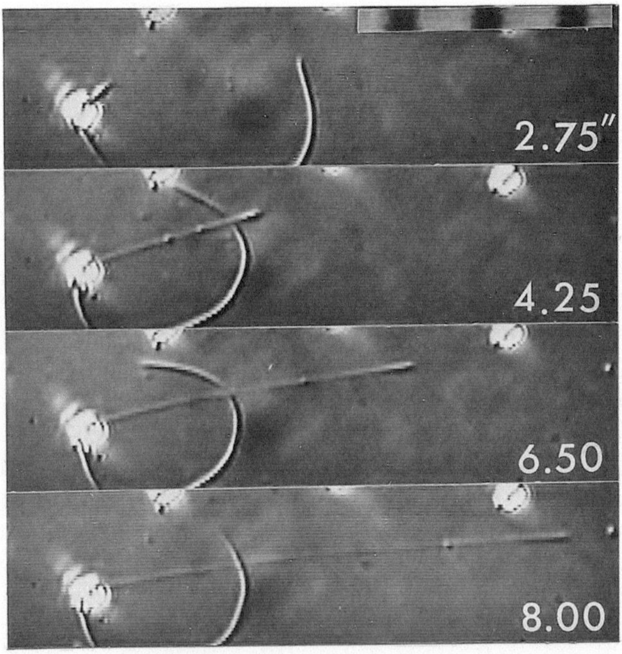

Fig. 4.26 DIC images of extending acrosomal process on *Thyone* sperm (Tilney LG, Inoué S, The acrosomal reaction of Thyone sperm. II. The kinetics and possible mechanism of acrosomal process elongation, J Cell Biol 93 (3): 820–827, 1982. [A 38])

With video-enhanced DIC, we could not only clearly see the thin, growing acrosomal process, which was only 65 nm in diameter, but also the formation, movement, and disappearance of tiny blebs along the length of the process. The curve plotting the elongation of the acrosomal process fits a diffusion-limited event, which buttresses the notion that unpolymerized actin molecules diffuse through the thin growing tube (i.e., the membrane of the acrosomal process) all the way to the tip of the acrosomal process, where they are polymerized onto growing ends of the actin filaments which support the growth of the process (Fig. 4.27; *Collected Works* A-37, A-38).

According to electron micrographs, the content of the acrosomal vesicle is emptied the moment the sperm 'pops' (i.e., when the forward end of the vesicle membrane fuses with the sperm membrane) before the acrosomal process starts growing. The video images, on the other hand, show that the acrosomal vesicle gradually becomes smaller as the acrosomal process elongates, but does not lose its density or concentration of its contents (Fig. 4.28). It is known that the acrosomal vesicle (that becomes everted as the actin filaments grow and its membrane is pushed from the back) contains the enzymes needed to penetrate the egg jelly and the protein needed to fuse the sperm cell membrane with the egg cell membrane. So it would only make sense for the vesicle contents not to be lost, as suggested by the EM work, but to

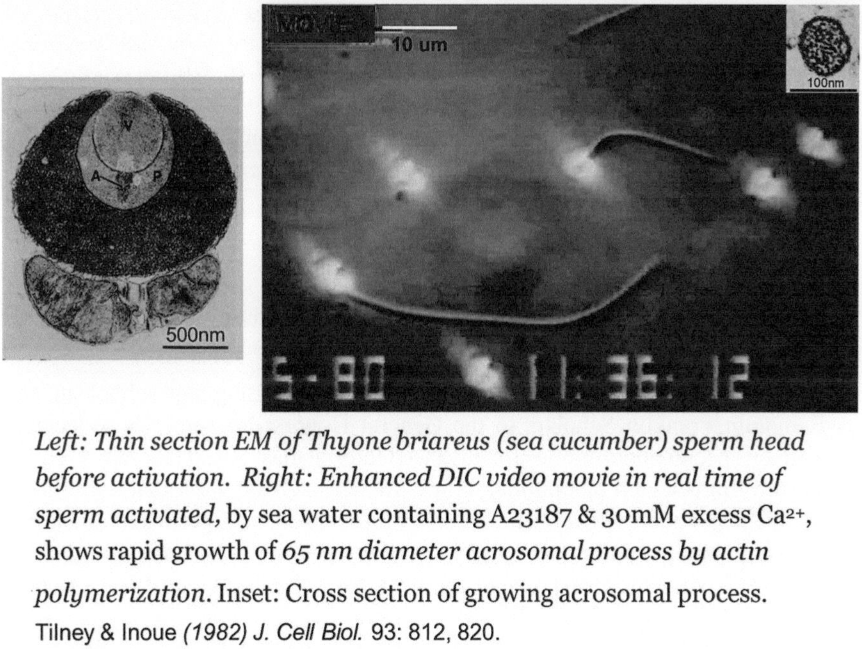

Fig. 4.27 EM of sperm head and DIC images of live sea cucumber sperm (Inoué S, Tilney LG, The acrosomal reaction of Thyone sperm. I. Changes in the sperm head visualized by high resolution video microscopy, J Cell Biol 93 (3): 812–819, 1982. [A 37])

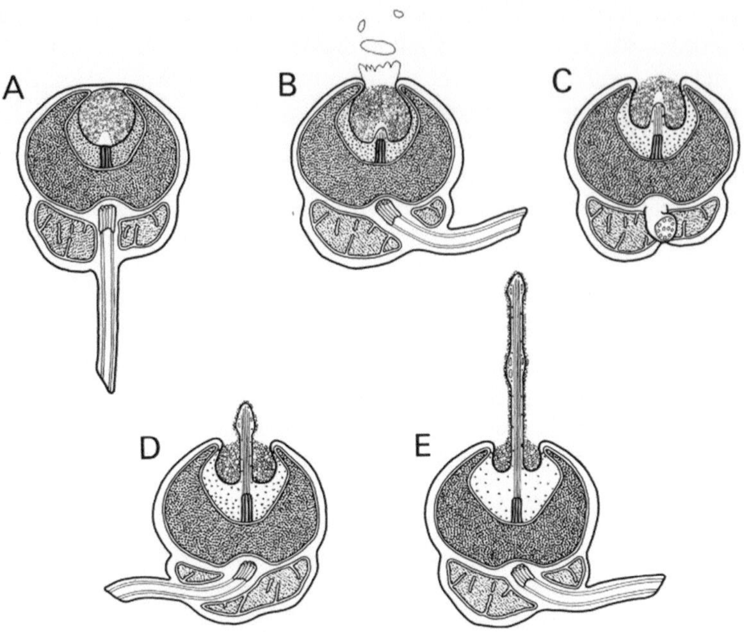

Fig. 4.28 Acrosomal process formation deduced from combined EM and live-cell DIC (Inoué S, Tilney LG, The acrosomal reaction of Thyone sperm. I. Changes in the sperm head visualized by high resolution video microscopy, J Cell Biol 93 (3): 812–819, 1982. [A 37])

gradually coat the outside surface of the extending acrosomal process (the everting membrane of the acrosomal vesicle) as the video DIC observation of the live specimen suggests.

Benjamin Franklin Scholars: Gradient of Anoxia

The University of Pennsylvania recruited especially promising high school students as Benjamin Franklin Scholars to the General Honors Program. Those students were given special opportunities for advanced tutorial classes and laboratory experience, many in research laboratories in the Medical School. I felt that these bright undergraduate students could benefit further if they were to experience some original research, patterned after our MBL Physiology post-course offerings.

So with cooperation and encouragement from Linda Wiedmann, Associate Director of the General Honors Program, we selected some half a dozen to a dozen Benjamin Franklin Scholars each year to spend three to 4 weeks at MBL in the spring. Doug Lutz, Dave Begg, and Dan Kiehart, who had completed their PhD in our lab, and Tim Otter, who got his degree with Ted Salmon, joined me in guiding those students at the MBL from 1982 to 1986.

During the first week of the course in Woods Hole each year, we introduced the General Honors students to experiments in physiology and embryology that involved living cells from local marine organisms or protists. At the same time, we introduced them to the use of various forms of light microscopy. During this week, they were to think of a project that they wanted to explore in further detail.

At the end of the first week, the whole class discussed the study proposed by each student, offering possible hints for their implementation. For the remaining 2 weeks or so, the students immersed themselves, first figuring out how to design and implement their experiments, and then proceeding with the actual observations and measurements. In the last 2 days, they wrote up their findings and presented the material to the whole group.

For many, if not most, of the students, planning and carrying out experiments on their own was a completely new experience. Up to that point, measuring the pH of solutions or counting the number of blood cells or chromosome types for Dr. So-and-so was what they thought research was all about. In Woods Hole, they were challenged to ask their own questions, to figure out practical ways of trying to answer the questions, to improve upon their approach, and then finally to make the needed measurement or observation, often flying by the seat of their pants. Yet we were most happily surprised by how quickly the students caught on and how effectively they managed to solve the problems that they encountered.

By the end of the session in 1982, some of the General Honors students had confirmed an unexpected observation, which I had made many years ago, and even managed to explain the unlikely mechanism at work. The observation that they confirmed is published in Article 36 in the *Collected Works of Shinya Inoué* (2008a). Fertilized eggs of star fish were placed as a single crowded layer in a tiny drop of seawater sandwiched between a microscope slide and coverslip. The micro-drop was surrounded by a donut-shaped air space, which in turn was surrounded by a hollowed-out piece of filter paper. Without contacting the micro-drop, this outer ring of filter paper, soaked with seawater, prevented evaporation of the micro-drop. The filter paper ring also kept the coverslip in a plane parallel to the slide and kept the two separated by the right distance. Finally, the preparation was sealed with "Valap" wax so that the chamber was isolated from the atmosphere outside (Fig. 4.29a).

My observation in 1986 went as follows: When the pancake-shaped micro-drop was smaller than 4 mm across, all the star fish eggs in the drop divided and developed normally, grew cilia in a day, and swam around in the tiny pool of seawater. But when the drop was any larger, say 6 or so mm across, only the eggs in the outer 2-mm rim of the drop would develop normally and those further in would cease to develop after a certain number of cell divisions. The group of cells at the middle of a large drop would not even divide once. What was totally unexpected in the embryos that stopped developing was that, ***within a single embryo***, only 100 μm in diameter, the cells facing the air space had divided once more than the next layer of cells, which in turn had gone through an extra division compared to the cells in the layer further away from the air space (Fig. 4.29b).

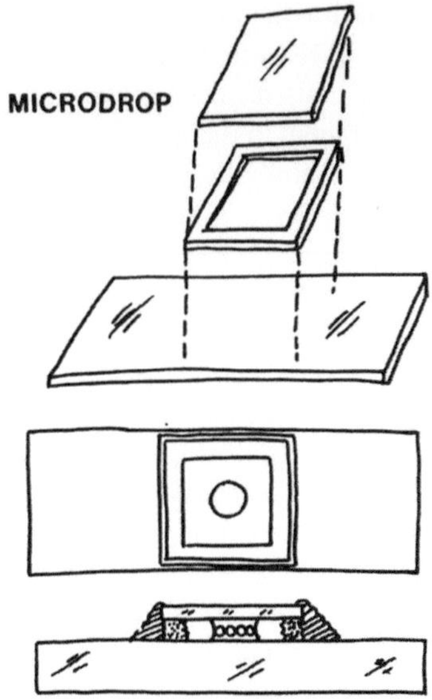

Fig. 4.29a Construction of microdrop chamber (height exaggerated) (Inoué S, Potrebic B, Brown CR, Lutz DA, Gradient of cleavage inhibition induced by limited diffusion of oxygen, J Cell Biol 95 (2): 136a, 1982. [A 36])

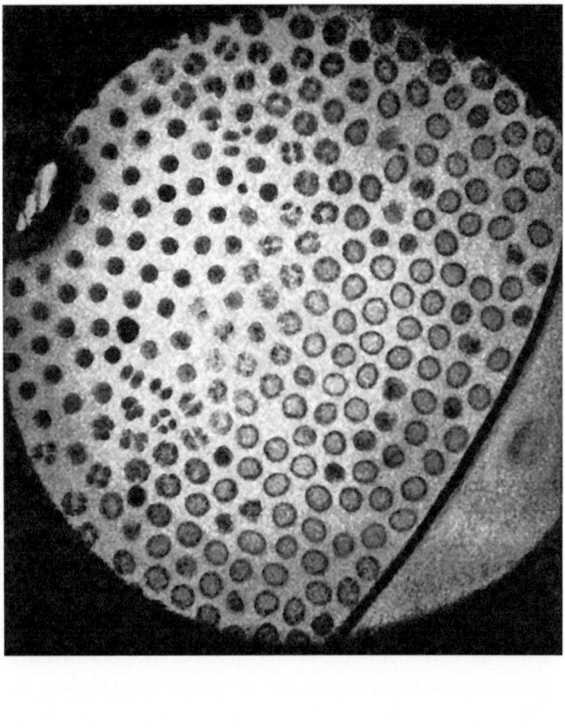

Fig. 4.29b Development of fertilized star fish eggs after a day in microdrop. Within ca. 2 mm from edge of microdrop, all embryos have developed to blastula stage; none of the eggs further than ca. 3.5 mm from edge of microdrop have divided at all; the cells in between have divided once, twice, three, four…times, depending on how far away it was located from the center of the microdrop (*Collected Works* A-36) (Inoué S, Potrebic B, Brown CR, Lutz DA, Gradient of cleavage inhibition induced by limited diffusion of oxygen, J Cell Biol 95 (2): 136a, 1982. [A 36])

Fig. 4.29c Anoxia-induced gradient of developmental inhibition in microdrop (Inoué S, Potrebic B, Brown CR, Lutz DA, Gradient of cleavage inhibition induced by limited diffusion of oxygen, J Cell Biol 95 (2): 136a, 1982. [A 36])

The two General Honors students who tackled this problem quickly confirmed my initial observation and took excellent photographs of the cells through the microscope (Fig. 4.29c). But the problem was how to explain the very sharp gradient of inhibition of cell division. Superficially, it looked as though oxygen could not reach the cells deep in the droplet so that only those near the edge of the drop could develop normally.

In fact, embryologists had known for years that in order to raise healthy embryos of sea urchin or star fish, one must avoid over-crowding so that not more than a few layers of eggs would settle at the bottom of the seawater in the beaker or finger bowl. The rational was that the eggs further down in the stacked layers would be deprived of the oxygen that they needed. But could the gradient of anoxia become so great that a whole cycle of cell division is inhibited across just a 30-μm (=100 μm/3) distance in the micro-drop (Fig. 4.29d)? How could eggs use up oxygen so rapidly compared to the very rapid diffusion rate of oxygen? It was too much of a puzzle so that I had even refrained from publishing my initial observations.

Then, Sonia Potrebic and Carol Brown proved that the cells were, indeed, responding to a gradient of anoxia (lack of oxygen) (*Collected Works* A-36). These students' experiments were so straightforward in concept, but required some technical feats that I didn't think were possible. They simply wanted to replace the air around the micro-drop with oxygen and watch the inhibition of cell division take place further in towards the middle of the drop.

So they drew out some heated polyethylene tubing to make them thin enough to squeeze through the gap formed by cutting the filter paper ring into two parts (Fig. 4.29e). The gap between the slide and coverslip was separated by a layer of moistened filter paper, just over 300-um thick. To my surprise, they managed to provide an inlet and outflow channel, with the preparation still sealed from the outer atmosphere, and showed exactly what they had expected: as the air surrounding the micro-drop was replaced with oxygen, the gradient moved further in, so that the

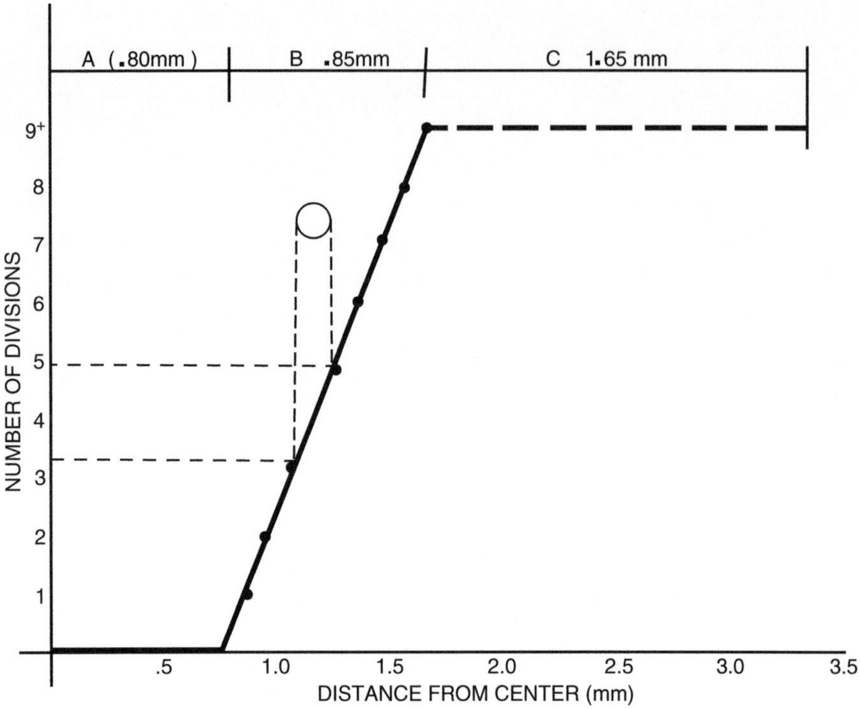

Fig. 4.29d Plot of anoxia-induced developmental gradient (Inoué S, Potrebic B, Brown CR, Lutz DA, Gradient of cleavage inhibition induced by limited diffusion of oxygen, J Cell Biol 95 (2): 136a, 1982. [A 36])

cells further in would divide more times, and eggs further in the micro-drop developed normally (Fig. 4.29f).

They also showed that the dye methylene blue {which turns colorless as the pO_2 (pressure of oxygen) drops} showed a gradient of anoxia consistent with the cell behavior. Finally, removing CO_2 from the preparation (by soaking the filter paper ring with KOH) did not change the gradient of cell division inhibition at all. So it was not that the CO_2 produced by the cells was inhibiting cell division, but that the cells did, indeed, use up the oxygen much faster than the oxygen could diffuse through a layer of several cells.

Perhaps that is why most of our own tissue cells are arranged in single layers, even in organs like liver and brain, and kept right next to the capillaries that supply oxygen and remove waste products. In any event, the findings by the General Honor ***undergraduate*** students were of sufficient interest, not only to be reported at the Annual Meetings of the MBL, but also at the 1982 Annual Meeting of the American Society for Cell Biology.

Given the appropriate chance and the stimulus, these bright young students had managed, in just a couple of weeks, to provide a clear explanation to a phenomenon that I myself had completely failed to explain!

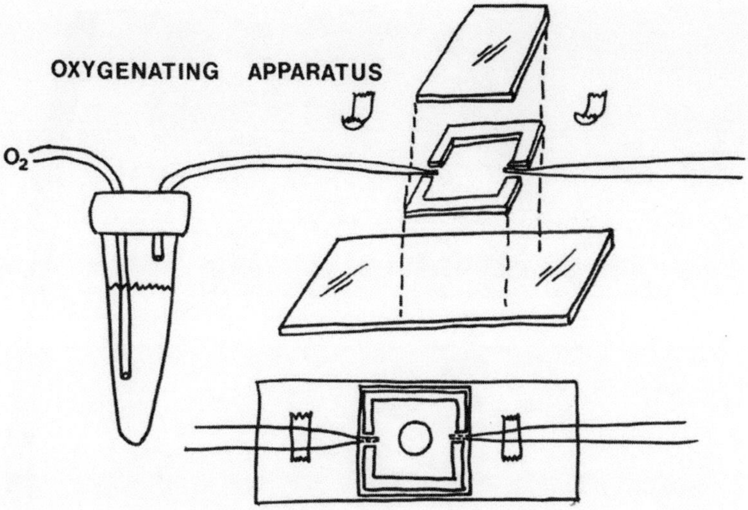

Fig. 4.29e Microdrop oxygenating slide devised by General Honors students (Inoué S, Potrebic B, Brown CR, Lutz DA, Gradient of cleavage inhibition induced by limited diffusion of oxygen, J Cell Biol 95 (2): 136a, 1982. [A 36])

Video Microscopy

By the time we had offered the MBL microscopy courses for 4 years, it was clear that a revolutionary new field was emerging. The combination of video and CCD cameras and digital computers with the microscope brought into view minute, dynamic worlds that had totally eluded detection, or even imagination, in the past. Advances in electronic image enhancement and information extraction, in turn, stimulated innovations and improvements in the optics and mechanical design of the light microscope itself. Classical barriers of light microscopy were being overcome, and people were beginning to explore new ways to use the light microscope undreamed of just a few years ago.

Thus, the time seemed ripe to summarize these new developments; introduce the potential users to the background that was needed to use and to contribute further to developments in this new field; show the actual ways and means; and explain how to properly interpret and appropriately share the image data. With encouragement from Kirk Jensen of Plenum Press and Mort Maser (who actually edited and proofed most of the chapters) and input from some colleagues in the MBL Microscopy Courses and many others, I tackled the writing of *Video Microscopy*. I turned to Bob Walters and Mike Berns at the University of California at Irvine to write the chapter on Digital Image Processing and Analysis and to Gordon Ellis of the University of Pennsylvania and Eric Hansen of Dartmouth College for two of the appendices.

Gathering the background material turned out to be an unexpectedly pleasant surprise, not only as a fresh learning experience, but especially owing to the willing-

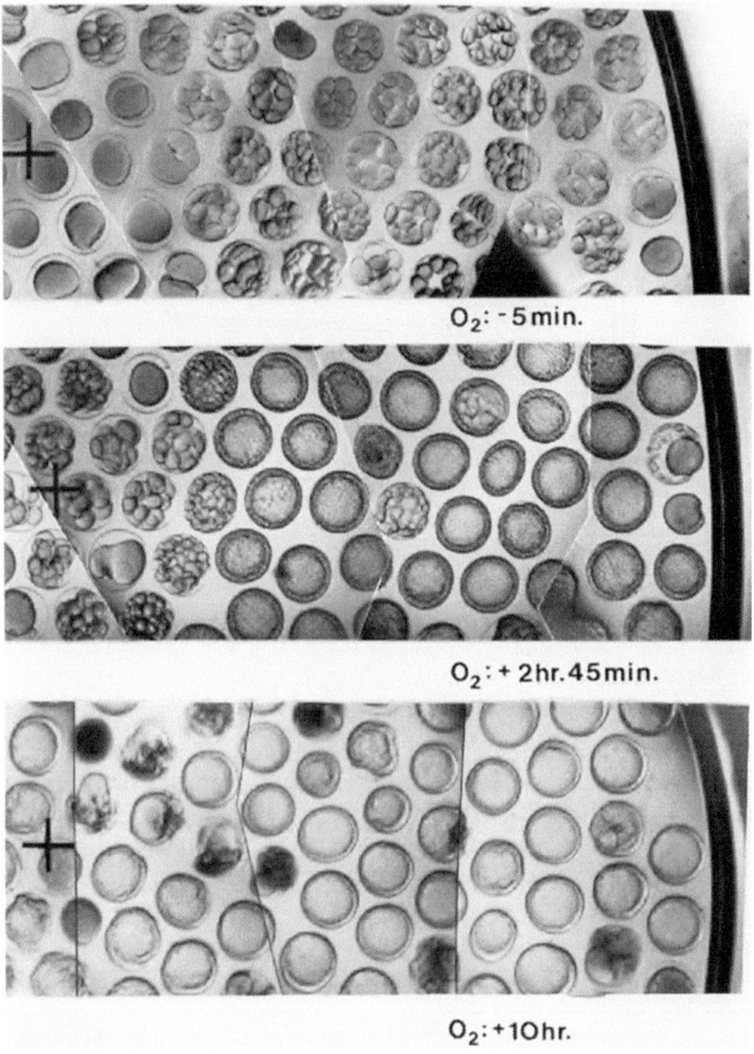

Fig. 4.29f Time sequence of embryo development in oxygenated microdrop. These sequential photos from the students' report show how replacement of the air surrounding the microdrop with oxygen (Fig. 4.29e) overcomes anoxic suppression of development (Figs. 4.29b, 4.29c, and 4.29d) of the star fish embryos

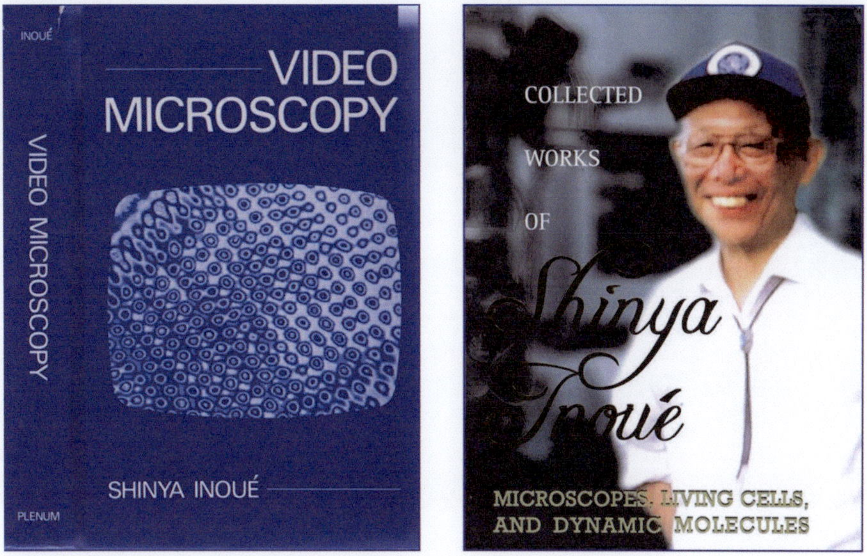

Fig. 4.30 *Left*: Author's *Video Microscopy*, Plenum Press, First Edition, with Appendix III, 1986 (Spanish Translation, 1993; Second Edition, 1997; Japanese translation, 2001). *Right*: *Collected Works* with appended DVD, World Scientific Publisher, 2008 (Courtesy SI)

ness of many manufacturers to share even unpublished information. I was also fortunate to have the able help of Jane Leighton who was Mort Maser's secretary (now Jane MacNeil, my executive assistant), who managed to convert 11 chapters, an appendix, and the glossary, all in my long-hand scribble, into a properly-worded typed manuscript. Thus, in April, 1986, Plenum came out with the first edition of *Video Microscopy* (Fig. 4.30, with the blue cover). Praising reviews appeared in *Nature* and other journals, and by 1989, the book was in its fourth printing. It was also translated into Spanish by Mario Burgos and collaborators and published in 1993 by the Inca Press of Argentina.

In 1997, Ken Spring of NIH and I co-authored the second edition, now called *Video Microscopy: the Fundamentals* (the green book). So much had happened, especially in advances in digital imaging and processing devices and technology, as well as in microscopy, that Ken contributed much new material, and most of the chapters were completely re-written. This edition was then translated into Japanese by Drs. Susumu Terakawa, Kouji Ichie, and Akira Watanabe and published in 2001 by Kyoritsu Shuppan Co. of Tokyo. During their translation, we responded to many questions raised by Drs. Ichie and Terakawa, so that the Japanese version is no doubt more precise in several places.

In all, Video Microscopy is widely cited as the major source reference, and I am delighted to see its continued contribution to the field. In order to stay within reasonable page limits, however, Appendix III in the first edition, which described the basics and some applications of polarization microscopy, could not be included in the second edition. This Appendix is reproduced in the *Collected Works of Shinya Inoué* as article A-40.

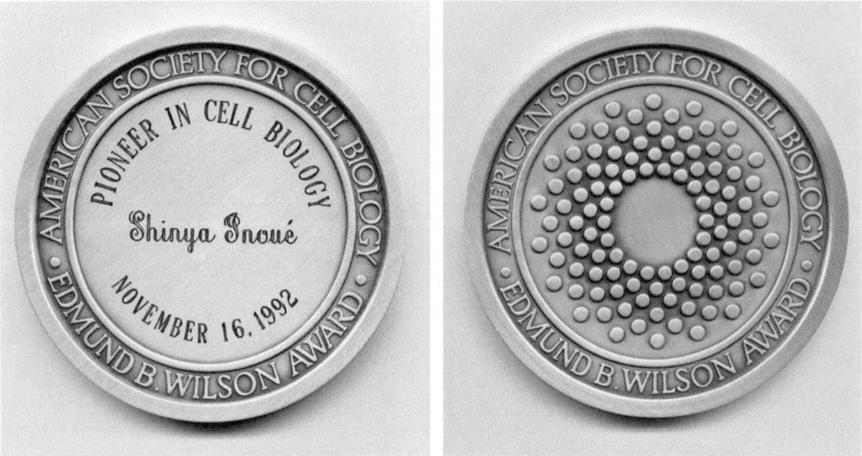

Fig. 4.31 E. B. Wilson award medal (*front and back*) (Courtesy SI)

E. B. Wilson Award

At the annual meeting of the American Society for Cell Biology in 1992, I was presented the prestigious E. B. Wilson award (Fig. 4.31).

I was given this award for: proving that spindle fibers were not artifacts of fixation, but were, in fact, present in living cells; proposing that molecular filaments (such as microtubules) pulled organelles to their target by de-polymerization, and pushed them by polymerization; and developing the microscopes which made such observations and insights possible. This was a special honor for me who had, since the 1940s, been studying mechanisms of mitosis using active living cells, while devising microscopes for revealing such mechanisms non-destructively.

In addition to displaying slides and movies revealing birefringent, dynamic cell organelles essential for mitosis and cellular motility generally, I chose this opportunity at the E. B. Wilson award to demonstrate stereo video images recently captured through our microscopes. Figure 4.32 shows the audience wearing Stereographics polarizing eyepieces which select the left- and right-tilted images. The use of polarizing glasses in order to obtain stereoscopic images by viewing the left- and right-tilted images, selectively through the appropriate eye, is illustrated in Fig. 1.15 (See *Collected Works* A-2, A-42).

Two Centennial

In April of 1987, the Misaki Marine Biological Station celebrated its 100th anniversary. Together with Jim Ebert, past president of MBL and President of the Carnegie Institution of Washington, I joined the festivities at Misaki and at Tokyo representing the MBL. Many others from the States who had worked at the MBL, including

Fig. 4.32 Author showing 3-D video images (Courtesy SI)

Dan Mazia, Charlie Metz, Ichiji Tasaki, Dave Epel, Vic Vacquier, Tom Humphreys, Gerry Schatten, and Rudy Raff also joined the international representations. Of the Japanese scientists, Katy Dan, Tokindo Okada, Masao Sugiyama, Yukio Hiramoto, Hikoichi Sakai, Hidemi Sato, and Kayo Okazaki, were among those affiliated with the MBL at one time or another.

In addition to the party at Misaki (Fig. 4.33), a 3-day symposium on *Perspectives in Marine Biology: Contributions to Cell and Developmental Biology* was held at the Sanjo-Kaikan Meeting Hall of Tokyo University. All of the individuals mentioned above contributed to the symposium. In his address at the meeting, the Crown Prince (who shortly thereafter succeeded Emperor Showa) implored investigators to save samples of organisms they were studying to make sure that they had properly identified the species; an astute comment from an experienced ichthyologist.

While delighted to join the celebrations at Misaki, many from the States were astonished that Tokyo University had not improved Katy's Lab since I had trained with him some 40 years before, after he had left the message "Last One to Go." While Toh-Dai could not refuse a scientist of Katy's distinction and contribution from continuing his work at the Station, they were not about to let a Professor from another University (Katy had been Professor and President of Toritsu, Tokyo Metropolitan University) even install an air conditioner needed to allow the urchin eggs under the microscope to develop normally without getting too hot. Katy had paid for the equipment, but for some reason, the Station management refused its installation!

Fig. 4.33 Katy & Masao Sugiyama at Misaki Lab's Centennial (Courtesy SI)

My month-long trip to Japan was sponsored by Motonori Hoshi of the Tokyo Institute of Technology (TIT), supported by the Japan Society for the Promotion of Science. So in addition to getting together with Katy, Kayo, and Tsuruta (my good friend from Musashi) and my four sisters, I had the opportunity to interact with the folks at TIT, in addition to Fumio Oosawa and his colleagues at Nagoya, Hidemi Sato and his associates at the Sugashima Marine Lab that he headed, Toshio Yanagida at Osaka, Takashi Shimizu in Sapporo, and Katsuji Rikukawa and colleagues at Nikon.

The trip also gave me a chance to get re-acquainted with Bjorn Afzelius, who was attending the symposium held in honor of Yukio Hiramoto's retirement from TIT. By then, Bjorn had clearly established the intriguing relationship between situs inversus, infertility, and flagellar anomaly. At Yukio's retirement party, the Institute's President surprised us by remarking how happy he was "that this ceremony finally gave him a chance to meet Mrs. Hiramoto in person!"

A year after the Misaki Lab's Centennial, and 2 years after that of the Naples Marine Lab, MBL celebrated her 100 years of contribution to science. One feature of the MBL celebration was a series of Centennial Friday Evening Lectures that was given during the summer months of 1988. The lectures, which were each dedicated to one or two prominent MBL pioneers, were edited by Bob Barlow, John Dowling, and Gerry Weissmann and published in 1993 by the Harvard University Press as *The Biological Century.* Those honored included: Jacques Loeb, Thomas H. Morgan, Alfred H. Sturtevant, Edwin G. Conklin, Frank R. Lillie, Albert Szent-Györgyi, Keith Porter, Winthrop J. V. Osterhout, Kenneth S. Cole, Alan H. Hodgkin, Selig Hecht, George Wald, H. Keffer Hartline, Stephen W. Kuffler, Charles O. Whitman, William M. Wheeler, Spencer F. Baird, Henry Bigelow, and Alfred C. Redfield.

I had the honor to discuss the contributions of Keith Porter, who had done much to shed light on the fine architecture of living cells with his innovative electron microscopy, and who had promoted the development of tissue culture as the President of the Tissue Culture Association. That also gave me an opportunity to discuss the complementary contributions of electron and light microscopy in our evolving understanding of the dynamic molecular and fine architectures of living cells.

Carl Zeiss contributed a historic calendar commemorating the MBL Centennial, and a street party in front of the Lillie Building was enjoyed by many, including the local public. I myself did not much enjoy such a party–I seem to be becoming more and more reclusive–but the occasion did provide an opportunity for reflecting on the differences between the two century-old institutions located in Misaki and Woods Hole.

With all its noted history and contributions to biology, Misaki clearly belonged to Toh-Dai. While I am told that the laboratory building was, in fact, renovated a year or so after we had visited, I doubt that the parochial atmosphere has seen much change. To be sure, the common room was open to many other colleges and universities that brought their students for field trips, and the collecting staff was most cooperative, but the bureaucracy of the government-run university seems to die hard.

With the Woods Hole experience in mind, and perhaps haven given up on reforming the Misaki Lab, Katy championed the development of the National Laboratory at Okazaki, where faculty members and young investigators from various universities in Japan and abroad could work side by side and interact with each other. To what extent Katy was satisfied with the outcome of his aspirations, I am not sure.

In contrast to Misaki, the MBL in Woods Hole was developed by a consortium of universities, colleges, and research institutes, and thrived by the interactions of investigators of all ages and background from around the world. Leaving their daily academic chores at their home institutions, immersing themselves in research, and sharing knowledge with each other for the whole summer, this community of scientists developed the foundations of basic biology year after year. Today, many of the MBL teaching programs still thrive, spawning innovators and leaders in many fields of basic biology.

I fervently hope that the development of the year-round programs, and growth of bureaucracy "in accommodations to the fiscal needs and government regulations," will not harm MBL's unique role in promoting innovations and discoveries in basic biology and interactions among scientists with different backgrounds and expertise. Such a failure could convert our special national and international treasure into a minor university or, worse yet, into just another medical research institute. The sea and the interface between water and land have much to teach us, not only about marine biology, but also about fundamental attributes of life, as testified by the continuing contributions from the MBL. Above all else, we must continue to emphasize our mission as the birthplace of knowledge for world-class investigators and students alike and not take comfort in simply contributing to its dissemination or its application.

Chapter 5
Further at MBL

Introduction

Cubuy, Canovanas, Puerto Rico

In 1988, a year before Hurricane Hugo devastated many of the houses and downed the ancient hard wood trees in the area, we had purchased "Toti's" cottage near the El Yunque Rain Forest on the northeast end of Puerto Rico (Fig. 5.1). Fortunately, Hugo's damage to our cottage was limited to the low-pitched roof above the bedrooms in the rear of the house. The many windows and glass sliding doors in the front, windward side of the storm, were protected by heavy gaged corrugated aluminum shutters.

Furthermore, we had escaped having to live with the month-long electrical power outage that followed the storm, since Sylvia and I had to fly back to Massachusetts 2 days before Hugo hit. What happened was that I had experienced the most severe chest pain, and Dr. Christman, a cardiologist in Falmouth (whom I had never met before), told me over the phone to fly back immediately for a complete checkup and treatment. The following week, three of my heart vessels were cleared out by balloon angioplasty at Brigham and Women's Hospital in Boston.

That was already 34 years ago, and the rain forest has long returned to a lush green. When we are here, away from the cold New England winter months, the tree frogs continue to serenade us with their songs, "coqui, coqui." Interestingly, the paper wasps, ubiquitous on the porch rafters before Hugo, have never reappeared. The city of San Juan is nearly an hour's drive away and her lights, as well as those of the planes coming in to land at the Munoz-Marin airport, entertained us each evening during supper. But my wife Sylvia and I seldom ventured into the city at night.

During our winter months at Cubuy, contact with my work at the MBL was limited to a phone call once a week from my secretary Jane MacNeil, followed by a

S. Inoué, *Pathways of a Cell Biologist*, Springer Biographies,
DOI 10.1007/978-981-10-0947-1_5

Fig. 5.1 Toti's cottage in Cubuy, PR (Courtesy SI)

Fedex package the next day. This peaceful abode, 3000 feet above the Caribbean Sea, was where I spent many mornings writing *Through Yet Another Eye.*

Being in the tropics, we were entertained by tame house lizards as well as colorful exotic plants, as well as occasionally by the 4-foot-long Puerto Rican Boa, a federally protected snake (Fig. 5.2).

One creature that attracted my attention was well hidden and, unless combated periodically, could upset our peaceful abode. It was the dry-wood termite. When they invaded the rafters or plywood ceiling or the softwood furniture, I would spot their presence by the pile of tiny ball-bearing-like droppings. The droppings helped me set up a plumb line to find their 1/16-inch diameter entrance plug that led to their hidden passage way. Then I would drill out their soft plug and squirt in some 'TermOut' or similar agent to flood and destroy the colony.

Some of the dry-wood termites also have a taste for books, although they are surprisingly selective. They seemed to like my *Video Microscopy* and especially the thick cover on *Mysteries of Ancient America* given me by our friend Charlie Ruiz in Cubuy. Being the critics that these termites seemed to be, I came across very few paperbacks that had the neat 1/16-inch round holes threaded through page after page, with one exception (out of some 20 paperbacks): Len Deighton's *Catch a Falling Spy.* Yet they didn't touch other mystery novel paperbacks by other authors printed by the same publisher.

Fig. 5.2 Chris and Phillipe Cano with a Puerto Rican Boa

Leaving aside such idle speculation, the termites do harbor a complex fauna of symbionts in their hind gut, without whom their hosts are unable to digest the cellulose they devour. Apparently the result of a very ancient symbiosis, the many hypermastigote protozoa that dwell there are obligatory anaerobes, that is, they cannot survive in the presence of oxygen. Thus the protozoa are killed, and as a consequence, the host even starves if the termite is exposed to a high atmosphere of oxygen. Furthermore, these symbionts (as well as their close relatives that dwell in the ancient wood-eating cockroaches) show all kinds of unusual cellular and biochemical features that exaggerate what is not readily seen in an average cell. Those nymphs, then, reveal intriguing and important tales about cells, which I have referred to earlier.

Looking Further Back

February 26, 1936, at our snow-covered home in Magome, we listened to the radio advising us to stay sheltered in a part of the house away from central Tokyo in order to avoid being hit by stray bullets from a military coup that had erupted that morning. With my parents and sisters away in Berlin, Germany, I was staying at our place in Magome with Grandfather Kiyomatsu and Aunt Mitsu, who were looking after me and my two cousins. Normally, I would have headed for my high school in Ueno on the local train which passes central Tokyo. The coup, known as the 2–26 Incident, was carried out by some 1500 junior officers of the Japanese Army over 3 days, and led to the assassination of the Prime Minister as well as the Ministers of the Treasury and Education. Using the guise of improved control of their ranks, the Army brass used the occasion to gain considerable power in Japanese politics.

Four years before this incident, on March 5, 1932, Katy's father Baron Takuma Dan, a graduate of MIT and head of the Mitsui Enterprises, was assassinated in Tokyo by a young right-wing Japanese "super-patriot." Katy himself was then a graduate student in Victor Heilbrunn's lab at the University of Pennsylvania.

I myself am now 91, and although way past the 20-year critical period following coronary angioplasties done in 1989 and 1990 and radiation treatment for prostate cancer in 1993 and 1994, I am finding my energy progressively dwindling. So instead of continuing to narrate my experiences and thoughts in too much further detail, in this final chapter, I shall highlight some recollections, events, and observations that I have so far not dwelt upon.

Father's Epitaph

This coming February 28th would have been mother's 113th birthday, had she not been taken from us on August 6, 1966. She had been operated and cared for at the Cancer Research Center in Tsukiji, near central Tokyo. During the last week before she passed away, she surprised me by insisting that she was not suffering from any pain, but instead by inquiring after how my children were doing. I should not have been taken by surprise because this was just mother's way as we shall soon see.

By the way, mother was born the same year as Sylvia's father, Kenneth W. McCandless. Kenneth was unscathed by the deadly Spanish flu, to whose sufferers he was repeatedly exposed during WW-I, and remained healthy in Walla Walla, Washington, until he passed away at the age of 102 (Fig. 5.3).

In the winter of 1966, my father had a chance to re-visit Eastern Europe and the United States on a 6-week trip, including 2 weeks at our home in Philadelphia. Not long after the trip, he finished the Epilog to Book 3 of his autobiography '*Sajo no Ashiato*' ('Foot Prints in Sand'). In this Epilog, he reminisces on events in the dozen years since he finished Book 3 and, in particular, on mother's service to the Empress Showa (Fig. 5.4a).

For my dear mother Hideko (Fig. 5.4b), known as 'mommy' to us five children and also to father, love and selfless service was second nature. Yet she was highly intelligent and discriminating, as reflected in her words, "I wish to live in truth" when proposed to by Kojiro, her husband-to-be. She not only chose right from wrong throughout her life, but led others by examples of her own action. Harsh words hardly ever left her mouth; instead her loving examples guided us children and others to live lives of service to others and to international understanding and peace.

Fig. 5.3 Sylvia's father Kenneth on his 100th birthday, together with Sylvia, her children, and her grandchildren

Fig. 5.4a Silver Bowl with Imperial Crest. This was given to Shinya's Mother by Emperor and Empress Showa

Disliking the dank, dark atmosphere of the ancestral Inoué graveyard in Hodogaya near Yokohama, father purchased a plot in the newly opened Kamakura 'Rei-en' (Memorial Park) in 1966. There he chose a burial site on a lawn-covered hillside with a view of Mt. Fuji on clear days, a site which very much appealed to him and his offsprings.

Fig. 5.4b Mother (Photo by Shinya on Magome house porch)

Father chose a granite gravestone and had engraved on it "*Michi Tohku-tomo, Ningen Kansei no Hi wo Inoru*" (Far Though the Path May Be, In Prayer for Man's Perfection) in his own handwriting (Fig. 5.5a). This epitaph without doubt also reflects mother's heart-felt prayers. Father had mother's ashes transferred from Hodogaya and placed in the grave together with some mementos of hers that he chose. On November 13, 1969, unfortunately the day after I had left him to return to the States, he passed away at the Tokyo Rosai Byo-in, where he was under the close care of Dr. Shunshiro Kondo. Father and mother now rest together under the granite headstone in Kamakura.

In an early chapter of these recollections, I described father's heart wrenching, but unsuccessful, effort in 1939 to prevent the signing of the Tri-axis Pact (which led eventually to Japan's attack on Pearl Harbor). Hopefully, though, father and mother are now resting in peace seeing how their children and grandchildren, in their own ways, are contributing to better understanding and harmony among people and nations.

In the last pages of his '*Sajo no Ashiato*,' on April 25, 1969, father expresses satisfaction and gratitude for the lives now led by his children and grandchildren, while wishing for their continued happiness and prosperity (Fig. 5.5b). He also dreamed of witnessing, before forever joining his wife Hideko with her sweet smile, my receiving an international prize that recognized my contribution to science.

Well, I was not accorded such honor during his lifetime, but I did receive several honors over the next quarter of century, culminating in the award of the International

Fig. 5.5a Headstone of Shinya's parents' grave in Kamakura Rei-En. (English translation added to photograph by Shinya)

Fig. 5.5b *Left*: Father's '*Sajo no Ashiato*.' *Right*: One of mother's drawings entitled 'Hagi'

Prize for Biology in 2003. Both father and mother would have been doubly proud of this Award from the Japan Society for the Promotion of Science, for it was a highly competitive international award and made in honor of Emperor Showa's life-long contributions to biology.

Last Quarter Century at MBL

International Prize for Biology

On March 13, 2003, I was met at the Raleigh-Durham airport by Ted Ueno, an amiable third-year graduate student at the University of North Carolina. As he drove me to the campus in Chapel Hill, where the dogwood and magnolia trees greeted us in full bloom, he told me how the graduate students in Biology were given a fund to choose and host their own speaker once each year.

Their choice of me as that year's speaker must have been influenced by their hearing quite a bit about our studies, since the Research Triangle area is active in cell biology research, and especially in cell division, led by Ted Salmon, Kerry Bloom, and others on the University of North Carolina campus and Bruce Nicklas, Dan Kiehart, and others at Duke University. In any event, my seminar presentation on *Polarized Light, Microscopy, and Living Cells*, which was my first attempt at using Power Point software, as well as the informal gatherings with the graduate students, went off smoothly and was enjoyed by all concerned.

Towards the end of my stay, Ted Salmon, at a party at his new home in Chapel Hill, asked if there was anything he could do for me, especially in view of my somewhat uncertain future at the MBL where my Federal grant funds had run out. My response: "I keep on getting a number of nomination forms for various prestigious awards; would he like to try nominating me for the International Prize for Biology? The area of selection for 2003 was cell biology; if awarded, it might help keep my laboratory going at the MBL."

Early on the morning of July 18, 2003, MotoNori Hoshi of Keio University called me excitedly from Tokyo to say that their committee had just chosen me for the 2003 International Prize for Biology, that there were 41 nominations from 14 countries, and that the parent committee at the Japan Society for the Promotion of Science (JSPS) would act officially on this nomination on September 11. A few days later, Toshio Yanagida from Osaka University called about a symposium associated with this Award to be held in Nara. While wonderful news and a great honor, Sylvia mentioned also what a good opportunity this would be to bring together family from both shores of the Pacific.

On November 22, 2003, Sylvia and I arrive at Narita Airport and then headed to Hotel New Otani in Tokyo, where we stayed until December 2. Between November 23 and 29, we attended my sisters Midori and Futaba's music concerts, my sisters' birthday parties, and visited Kamakura Rei-en. Our children and grandchildren from Santa Fe, Seattle, and New Hope, Pennsylvania, arrive in Tokyo, and Sylvia and I attended a dinner hosted by Nikon.

Then, on November 30, some 40 friends and family gathered at the Tokyo American Club (Fig. 5.6). My youngest sister Futaba Inoué Koike's piano performance of Fauré's pieces preceded my Power Point presentation on *Synopsis of 2003 International Prize for Biology* including some snapshots of our family in the States and in Japan, which was followed by my nephew Tsukahara Kota's piano jazz rendition of *Over the Rainbow*.

Fig. 5.6 March 2010 family gathering in Tokyo

The Ceremony

On December 1, 2003, at 9:40 am, Sylvia and I were met at our hotel by Ms. Yuko Furukawa from JSPS and escorted to the Japan Academy in Ueno, Tokyo. In the recipient waiting room, Sylvia and I were photographed together, with me holding the Imperial gift of the silver vase embossed with a gold chrysanthemum crest (Fig. 5.7).

Then we were introduced to JSPS representatives and other dignitaries. This was followed by a private meeting with their Majesties the Emperor and Empress, whereupon her Majesty the Empress conversed with Sylvia in fluent English. I learned from his Majesty that he was in high school and not grade school at the time of his visit to Misaki where we had first met, and that he was intrigued with the electric ray, not electric eel, as I had remembered. (This information led to an impromptu revision of my soon-to-be-delivered acceptance address.)

At 11:20 am there was a practice session of the Ceremony that included the family members (all 5 of our children, 3 of their spouses, and all 4 of my sisters) taking their seats.

Nine minutes later, their Majesties the Emperor and Empress enter the auditorium and Ceremony proceeded (Figs. 5.8a and 5.8b; see also Appendix II, Address by his Majesty the Emperor, and Appendix III, Acceptance Address by Shinya Inoué). After the Ceremony, a mixer and buffet was held in the presence of their Majesties the Emperor and Empress for all those attending the Ceremony (Fig. 5.9), and later that evening, Sylvia and I attended a dinner at 'Tai-kan En' hosted by JSPS.

Fig. 5.7 Sylvia and Shinya with Imperial vase

Fig. 5.8a Award ceremony. Shinya Inoué receiving the ceremonial award and medal from Dr. Saburo Nagakura, Chairman of the Selection Committee

The next day, Sylvia and I, together with our oldest son Jon with his children Sean and Lauren, and our second son Christopher with his wife Caroline and daughter Katie, travel by Shinkansen express train 'Nozomi' ('Hope' or

Fig. 5.8b International prize medal

Fig. 5.9 My sisters, Futaba, Midori, and Akiko are introduced to their Majesties

'Wish') to Kyoto. From there, I headed off by myself to Nara to attend the Commemorative Symposium.

The Symposium, entitled *Cellular and Molecular Dynamics Revealed by Advanced Light Microscopy*, was held on December 3 and 4 and was organized by

Fig. 5.10 Nara symposium speakers

Toshio Yanagida of Osaka University. There were 8 speakers from the United States, 4 from Europe, and 11 from Japan (Fig. 5.10; Dell and Vale 2004).

Following the Symposium, I joined my family in Kyoto for 3 days of sightseeing and recreation. Then Sylvia and I flew back via Los Angeles to Riverside, California, where we visit with Kenneth and Ruth Cooper (Fig. 5.11) before returning to Falmouth on December 10.

All told, it was an exciting and memorable experience, not only because of the Award Ceremony, but also because of the pride and pleasure of bringing nearly the entire Inoué family from both coasts of the Pacific together.

Dan Fellowship

In 1970, 3 years before Katy's retirement as President of Tokyo Metropolitan University, Jean Dan started commuting to the newly opened Marine Laboratory in Tateyama on Boh-soh Peninsula across Tokyo Bay from Misaki. The new laboratory belonged to the Ochanomizu Women's College, where Jean was teaching. Because the countryside and weather were especially appealing to Jean, who occasionally suffered from asthma attacks, she soon purchased some land on which she planned to build a home and grow fresh produce. Becoming friendly with the local farmers and carpenters, her dream came into being some 3 years later. There, in

Fig. 5.11 Kenneth and Ruth Cooper in Riverside, CA

addition to farming to her heart's content, she cared for one of her three grandchildren, whose mother had died in a car accident in Brazil.

Late in 1978, while Katy too was using the Laboratory at Tateyama, Jean's best friend Suzie Steinbach from Woods Hole came for a ten-day visit. But totally unexpectedly and much to everyone's sorrow, on November 13, 7 days after Suzie's arrival, Jean succumbed to a fatal asthma attack and passed away. The funeral procession, joined by the local farmers along the levees of the rice paddies, is poetically pictured in Katy's *Uni tó Kataru* (Dialog with Sea Urchins) (Fig. 5.12, left).

The following year, Jean and Katy's oldest son Teruki and his wife Alice arrived in Woods Hole where the MBL Director Paul Gross, who was also a former graduate student with Victor Heilbrunn at Penn, made a Lab boat available to them. From the boat, some of Jean's ashes were scattered in the Hole beyond Nobska Point watched over by Suzie, Paul, Jean's children, her granddaughter, and others (Fig. 5.13).

In 1980, Katy himself came to Woods Hole as the MBL Lillie Fellow, and while he stayed at the Steinbach's, we decided to inaugurate the Dan Fellowship in Jean's memory and to commemorate the pioneering contributions of both Jean and Katy. The funds would be used to encourage further exchange of young biologists between Japan and the United States.

In 1981, the first Dan Fellow Melanie Pratt, who had worked with Ray Stephens at the MBL, spent a summer at the Biochemistry Department of Tokyo University headed by Hikoichi Sakai. Hiko was one of Katy's earliest graduate students at

Fig. 5.12 Books by Katsuma Dan and re Jean Clark Dan. *Left*: Katsuma Dan, 1987, *Uni tó Kataru* (Dialog with Sea Urchins) in Japanese. *Middle*: Y. Hagiwara ed., 1982, *Azur Sister Jean Clark Dan* (Collected memories of Jean, her vita, publication list, and two articles by Jean) some in English and some in Japanese. *Right*: Kyoko Kato, 1980, *Nagisa nó Uta* (Song by the Sea; Life History of a Female Biologist) in Japanese

Fig. 5.13 Memorial for Jean Clark Dan (Courtesy SI)

Tokyo Metropolitan University. There, Melanie shared their common interests in the biochemical characterization of tubulin and microtubules. In turn, Yuichiro Tanaka, who had worked with Hidemi Sato at Sugashima, joined us at the MBL from Japan.

Using a low concentration of Cytochalasin-D, an inhibitor of actin polymerization, Yuichiro showed how the cleavage furrow in *Arbacia* eggs could be made to retract temporarily and reversibly. During the rapid retraction, by tracing the movements of the echinochrome granules embedded in the egg cortex, he found that that the furrow retracted like a rubber band that snapped; in other words, the granules moved back rapidly along the cleavage furrow.

What followed in the reformation of the furrow was, however, totally unexpected. Instead of those same granules being led into the recovering furrow, other granules located at some distance away from the furrow were drawn towards, and incorporated into, the furrow (see Figs. 7 & 8 in *Collected Works* A-51). In other words, new cortical regions (and no doubt their actin molecules) appear to migrate at right angles to the furrow and then generate the reforming furrow; an intriguing insight into the mechanism of contractile ring formation.

From the United States, Melanie Pratt was followed in succeeding years by Mary Porter, Carl Johnson, Ron Vale, and Doug Koshland to various laboratories in Japan, while we at the MBL hosted Yukihisa Hamaguchi from the Tokyo Institute of Technology in 1982. In the section *Biophysical Cytology* in Part 4 of this article, I already described the exciting work by Hamaguchi et al. regarding the displacement of the *Chaetopterus* meiotic spindle by micro-manipulation followed by its reorientation and re-attachment at the animal pole (*Collected Works* A-45).

These Dan Fellows were not only "promising young investigators who have proven themselves in research" as Katy had desired, but they are leaving clear imprints on the progress of cellular and developmental biology, as well as on international cooperation and understanding. The Japanese Committee of the Fellowship followed Katy's wishes and terminated their Program in 2000, while the U.S. Fellowship remains as an endowed fund at the MBL that supports the Jean and Katsuma Dan Lectureship which brings together students and faculty of the MBL Physiology and Embryology courses.

As a long-time member of the Laboratory and one who had been involved intimately with her two microscopy courses, I was able to help initiate two other summer Fellowships at the MBL. I believe it was in 1991 that Lee Shuett at Nikon, Inc., whose members had been major contributors to our microscopy courses, asked me whether Nikon could do something special for the Laboratory. By then Carl Zeiss had already established a year-round support center at the MBL, manned by such outstanding and helpful experts as Phil Presley followed by Rudi Rottenfusser with support from his lovely wife Hanna. Fortunately Lee and his Japanese parent company enthusiastically took up my suggestion and agreed to set up a generously funded Nikon summer Fellowship for MBL. Then around 1999, I was able to have our son Ted, who headed Universal Imaging Corporation (UIC), and Jeff Stuckey the CEO of UIC (another major participant and supporter of the MBL microscopy and other courses) also provide significant funds for summer MBL Fellowships.

From 1992 to 2003, the successive Nikon Fellows were: Andreas Stemmer, Vladimir Gelfand, Andrew Murray, John Murray, Jeff Lichtman, Gregg Gundersen, James Zheng, Ted Salmon, Peter Saggau, Fred Chang, and Alexey Khodjakov. The UIC Fellows from 2000 to 2003 were: Tim Mitchison, Prem Ponka, Fred Chang,

Fig. 5.14 Rudolf Oldenbourg (Courtesy R Oldenbourg, Nature 381, 1996)

and Gregg Gundersen in succession. The list of these outstanding individuals suggests the major impact that these fellowships continue to have on the summer science and educational program at MBL. At the same time, we trust that they are likewise benefiting the sponsoring companies.

Architectural Dynamics in Living Cells (ADLC) Program

On April 27, 1988, I had the honor of receiving the *Lewis S. Rosenstiel Award for Distinguished Work in Basic Medical Research* from Brandeis University. Attending my lecture on that occasion was Rudolf Oldenbourg (Fig. 5.14), who had hailed from Germany and was an Assistant Professor in the Physics Department at Brandeis University since 1984. Somehow he was so intrigued with my presentation that he decided to join our laboratory and became Assistant Scientist at the MBL from 1989 to 1993. (Since then, he progressed through Associate Scientist and has become Senior Scientist since 2002). With the research personnel expanding, our group became called the Architectural Dynamics in Living Cells (ADLC) Program in 1992.

One of Rudolf's earliest contributions at MBL was to explain 'edge birefringence,' which I had first described in 1959 while examining the detailed distribution of birefringent band patterns in cross-striated muscle (Inoué 1959). Edge birefringence, an optical phenomenon unrelated to molecular anisotropy, is found at the sharp boundary, or microscopic interface, of any two regions where the refractive indexes on both sides of the boundary are different. The material on the two sides can both be solids or liquids, or the interface can be between two materials of different phases, e.g., between a solid and liquid or gas, or between a gas and liquid (such as an air bubble in water).

Whatever the state of the media, edge birefringence appears as a thin double layer, with the slow axis oriented parallel to the interface on the higher refractive index side and perpendicular to the interface on the lower refractive index side. When the refractive index of one of the media is changed, edge birefringence disappears when the refractive indexes on both sides of the interface become equal (leaving only the superimposed intrinsic birefringence which is due to intrinsic anisotropy of the underlying molecules). When the refractive index of the lower index material is raised further, the sign of the two birefringent 'layers' reverse as the lower index side now gains a higher index than the other side.

When the refractive index of a medium imbibing an object is varied, the behavior of birefringence due to edges, or interfaces, is distinct from that observed for material whose birefringence originates from intra-molecular or inter-molecular anisotropy. When the birefringence originates solely from intra-molecular anisotropy (known as intrinsic birefringence), the retardation does not change as the refractive index of the imbibing medium is altered.

When the birefringence originates from inter-molecular or fine-structural anisotropy (known as form or textural birefringence), the retardation changes parabolically as the refractive index of the imbibing medium is altered. At the matched index point (where the refractive index of the imbibing medium matches that of the fine structural rods or plates), the form birefringence disappears, just as in edge birefringence, and only the intrinsic birefringence remains. Above the match point, the form birefringence reappears but with the same sign as in the lower index immersion medium (*Collected Works* A-28). Thus, with form birefringence, the response to high index imbibing media is the opposite from what is seen with edge birefringence.

Rudolf provided a concise explanation of edge birefringence based on model experiments using KCl crystals (which are isotropic) and geometrical optical analyses (Oldenbourg 1991). While Rudolf's work provided important insight, I hope the phenomenon will be examined further from a wave optical viewpoint, or as a diffraction optical phenomenon. By doing so, I think we should arrive at a unified theory that predicts the exact anisotropic diffraction patterns arising from edge birefringence and from thin objects exhibiting form birefringence. Since such analyses should also uncover the influence of the wavelength of light and numerical aperture of the optics involved, one may well arrive at a non-invasive way (i.e., without the need to change the imbibing medium) of distinguishing the different sources of the birefringence observed by analyzing the wavelength and numerical aperture dependence of retardation.

After his analysis of edge birefringence, Rudolf developed what is today known as the LC-PolScope, a polarizing microscope incorporating certain electronic com-

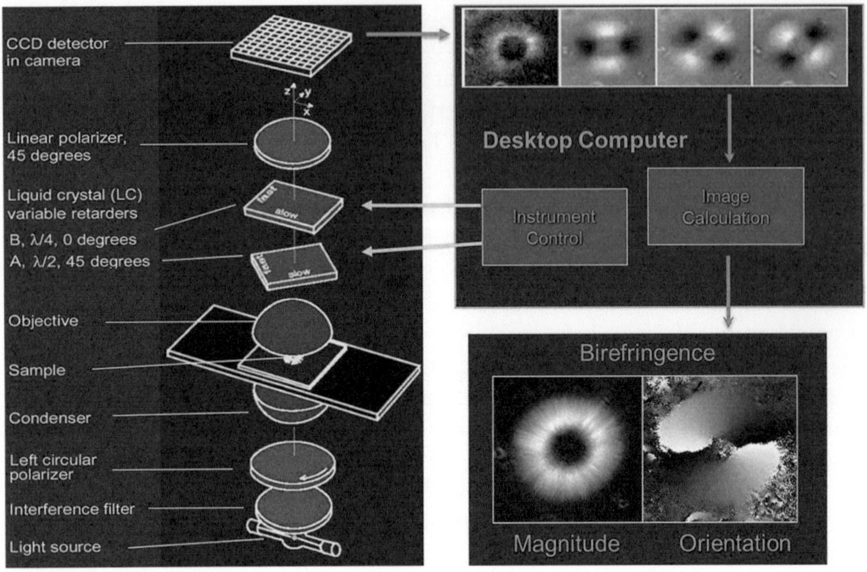

Fig. 5.15 Oldenburg's LC-PolScope

ponents and integrated with a digital computer (Oldenbourg and Mei 1995; Oldenbourg 1996). Without the need to manually manipulate a compensator or change the orientation of the specimen, Rudolf's system generates images (in fact, a map or maps) of the distribution of birefringence retardance and slow axis orientation of every pixel in the field of view in a few seconds (Fig. 5.15).

In comparison to the laborious analyses, which took three of us nearly 3 years and measured the distribution of retardance and azimuth orientation in three sperm heads of a cave cricket (*Collected Works* A-24), Rudolf's new system is indeed of the rocket age. In addition, despite my initial concern about image degradation by the liquid-crystal compensators used in the optical path of the LC-PolScope, the image quality turns out to be superb, as also is the sensitivity for detecting and measuring very weak birefringence, independent of the slow axis orientation of the specimen (Fig. 5.16). [However, the image generated by the LC PolScope does not quite supersede the best of the images generated using a high-extinction rectified system and Brace Koehler compensator. This is because within *each* image captured using a rectified system, the orientation-dependent black and white images do quantitatively reflect the retardance value and orientation of the slow axis for each image point in the specimen.]

In the development and application of the LC-PolScope and related projects, Rudolf has been associated with several individuals, including Jim LaFountain from University at Buffalo, Susan Gerbi from Brown University, and Amy Gladfelter from Dartmouth College, plus Grant Harris, an enthusiastic computer programmer. Rudolf's association with David Keefe and Lin Liu of MBL and the Infertility Clinic in Providence, RI, has resulted in a commercial version of the LC-PolScope,

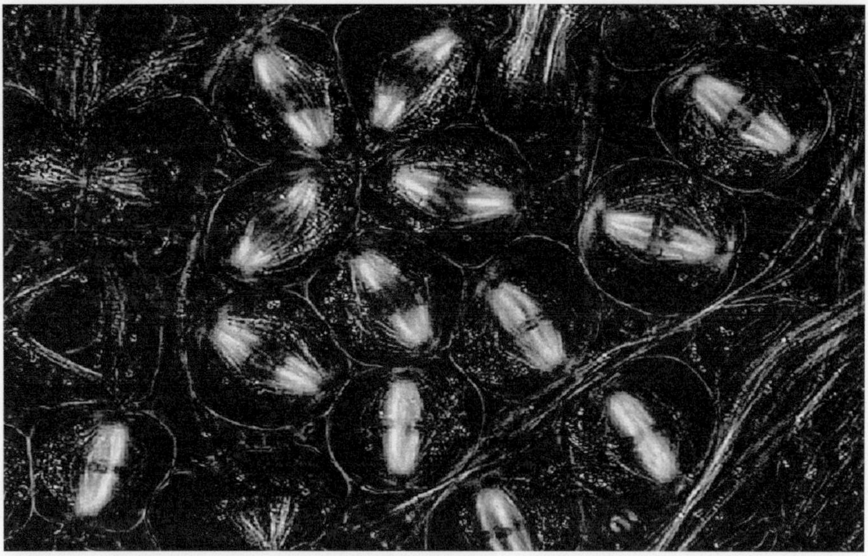

Fig. 5.16 LC-Pol image. Birefringence of meiosis-II spindle fibers in live dividing spermatocytes of crane fly. NB! orientation independence (LaFountain and Oldenbourg 2002 unpublished image. Courtesy of James LaFountain, University of Buffalo and Rudolf Oldenbourg, MBL Woods Hole, MA)

which is now produced by the Cambridge Research Institute in Cambridge, MA, and widely used by fertility clinics.

Mike Shribak (Fig. 5.17), an ingenious optical engineer from Belarus, was earlier associated with Rudolf, but is now an independent investigator in our ADLC Program. In the past few years, he has added the OI-DIC Scope [Orientation Independent Differential Interference Microscope (Fig. 5.18)] to his list of inventions. This new optical system captures orientation independent DIC images of live cells with unprecedented clarity (Fig. 5.19).

At the same time, the system can provide color-coded images representing interference gradient as well as dry mass distribution, or optical path differences (Fig. 5.20, *Color version on p. xii.*).

Other Collaborations

In the last quarter century at the ADLC Program, I myself have had extensive collaboration with several microscope and other instrument manufacturers as well as with visiting scientists. The former included: Hiroshi Takahashi, N. Okada, and Keisuke Suzuki (Olympus); M. Iida and H. Ichie (HPK); Katsuji Rikukawa, Kumiko Matsui, and Rieko Arimoto (Nikon); Ted Inoué, Dan Green, et al. (UIC); and others (Figs. 5.21a and 5.21b).

Fig. 5.17 Michael Shribak (Courtesy M Shribak)

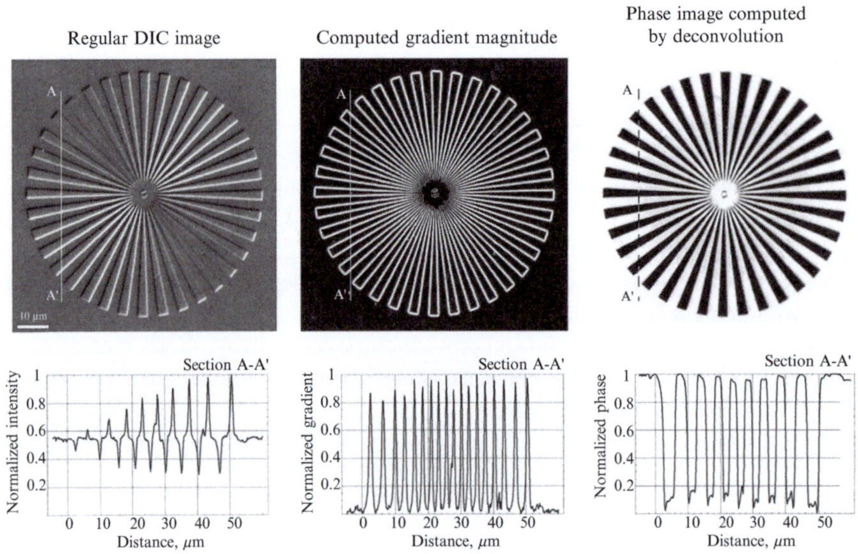

Fig. 5.18 Test images obtained with OI-DIC Scope (Shribak M, Inoué S, Orientation-independent differential interference contrast microscopy, App Opt 45 (3): 460–469, 2006. [A 74])

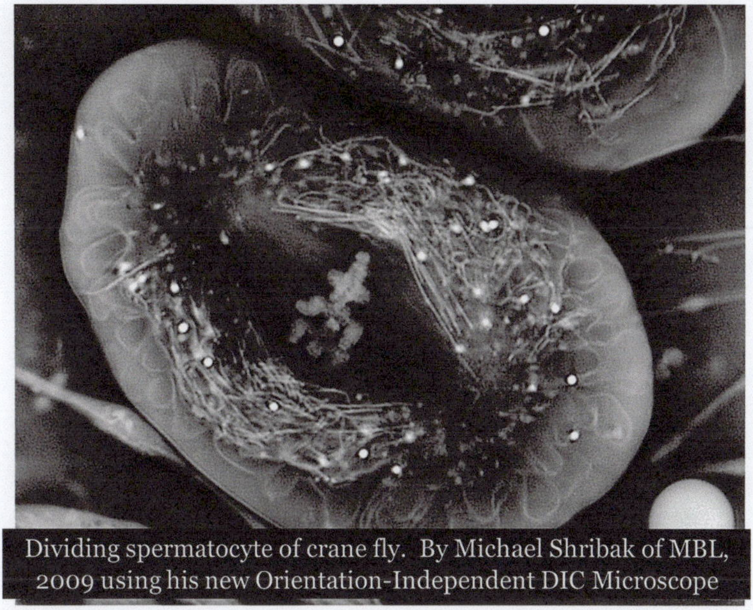

Fig. 5.19 OI-DIC close-up image of crane fly spermatocyte (Courtesy M Shribak, 2009)

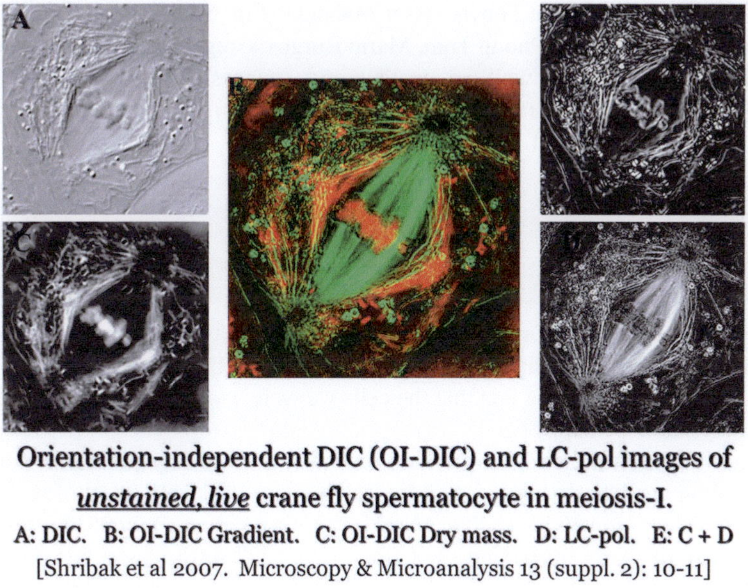

Fig. 5.20 OI-DIC and LC-Pol images (Courtesy M Shribak, 2007)

Fig. 5.21a ADLC program collaborators. *L.* to *R.*: Kumiko Matsui (Nikon), Katsuji Rikukawa (Nikon), Bob and Corey Knudson, Shinya and Ted Inoué, and Makoto Goda (Courtesy MBL)

The latter include Fumio Oosawa, Andrew Szent- Györgyi (Fig. 3.36), Yoshinori Fujiyoshi, Jean & Colette Febvre, Issei Mabuchi (Fig. 5.22a), Yoshio Fukui, Ted Salmon, Tim Mitchison, Phong Tran, Mario Burgos, Gregg Gundersen, and several others (Fig. 5.22b), as well as postdoctoral trainees such as Makoto Goda (Fig. 5.21a), H. Terada (from Hamamatsu Photonics), and a graduate student Danielle France from MIT.

'Happy Cells'

Over the years, several of my students and coworkers have heard me say "There is no point working with unhappy cells." What did I mean by that and how did I come to be so sensitive about the fact that the cells we were working with (actually learning from) had to be happy?

What truly left a lasting impression on me was Katy (my professor Katsuma Dan) telling me, when we were studying cell division or development of sea urchin and jelly fish eggs in 1947–1948 at Misaki, "That's no good. Only 95 % of the eggs got fertilized!" One's initial naive reaction might be, "Maybe just a few percent of the eggs were either immature or past their prime, or the sperm were not quite able to fertilize all of the eggs," or that something else was subtly wrong.

But whatever the problem was, Katy would insist that the batch of cells was not to be trusted as being normal and that we should start the observation or experiment all over again. Sometimes it was a matter of choosing another female for the eggs,

Fig. 5.21b *L.* to *R.* Keisuke Suzuki, Rudolf Oldenbourg, Bob Knudson, Rieko Arimoto, Shinya Inoué, and Jane MacNeil

Fig. 5.22a Issei Mabuchi and Shinya (Courtesy MBL)

Fig. 5.22b MBL imaging group (2010). From *top Left* to *Right*: Jason Swedlow, Conly Rieder, Rudolf Oldenbourg, Tim Mitchison, Gaudenz Danuser, Jim LaFountain, Ron Vale, Gary Borisy, Mike Shribak, Ted Salmon, Kip Sluder, and Shinya Inoué (Courtesy MBL)

other times a different male or fresh sperm, other times there was something wrong with the seawater, or maybe the room temperature had become too high. Whatever the reason, one had to figure out the remedy, perhaps wait until the next day, and avoid describing the untrustworthy events taking place in unhappy cells.

Around 1943, even before coming to Misaki to work with Katy, I had been exposed (as a student at Toh-Dai) to Dr. Tamiya's plant physiology course where we discussed osmotic pressure and plasmolysis of plant cells, and how transpiration of water and gas exchange from leaves are accelerated by the presence of the many tiny pores in the plant's epidermis (water evaporates faster from a dish with a cover containing many small holes than from an uncovered dish!). And in Dr. Kamada's animal physiology course, we experimented with the osmotic-pressure-dependent pulsation frequency of contractile vacuole in marine ciliates, as well as with the sodium/potassium and calcium/magnesium ion balance on the longevity of excised frog heart (phenomena that were initially discovered by Ringer). I also had been sensitized to the contracture that even the tiniest wound would induce in muscle cells (*Collected Works* A-3), but the need to listen to, or pay close attention to, the subtle signals ***uttered by the cells themselves***, was a lesson that truly sunk in when I worked with Katy.

So how did such a lesson affect my studies of living cells? And what surprising insults do the cells actually not mind if you let them tell us of their response? I already mentioned how the drop in fertilization rates at Friday Harbor told us that the running seawater was coming through the old pump with metal pipes (and thus contaminated by trace amounts of heavy metal ions) instead of through the new all Pyrex system, and how for the same reason we would go out in the old days to Nobska Point to collect fresh seawater for use at the MBL. Even with good seawater and well fertilized eggs, the abnormal shape or birefringence pattern of the spindle would tell us right away that the cells were feeling too hot or that the seawater was evaporating from the unsealed edge of the coverslip.

Organisms in Puget Sound, where the Friday Harbor Labs are located, live in seawater whose temperature seldom fluctuates from 11 ° C. In order to observe cell division and spindle behavior in cells from those organisms, I was able to keep them happily developing by perfusing seawater from the Sound through a handmade temperature-controlled slide in which the coolant seawater flowed through a channel separated from the cells by a single glass coverslip (*Collected Works* A-22, Figs. 3.13a and 3.13b). Those embryos would grow happily for over a day in a micro-drop chamber, so long as the prep was sealed properly and the cells were not too crowded in a large micro-drop and suffering from anoxia (*Collected Works* A-36).

During my thesis research at Princeton and at Woods Hole, the pollen mother cells of the Easter lily, and oocytes of the parchment worm *Chaetopterus*, were to play major roles in my being able to visualize the spindle fibers and fibrils and their behavior during cell division (*Collected Works* A-10, A-11, A-13). Yet both of these cell types were so full of birefringent granules that they had to be centrifuged to stratify the granules and, in the case of *Chaetopterus* oocytes, even fragmented by gradient centrifugation in order to obtain clear partial cells so that the weakly birefringent spindle was clearly visible with polarized light microscopy. Despite those seemingly drastic treatments, the centrifugally-cleared cells or fragments would tell me they did not mind by continuing their successive two divisions (in the case of pollen mother cells of the Easter lily), or by allowing meiosis to proceed when the fragments were fertilized (in the case of *Chaetopterus* oocytes). That experience also made me feel that it was worth designing and fabricating a centrifuge polarizing microscope (*Collected Works* A-66, A-67, A-68, A-75).

While excessively high temperature does induce irreversible damage, cells can tolerate surprisingly low temperature, and reversibly so. This property was important for showing how the spindle fibrils (microtubules) could be depolymerized and pull chromosomes and centrosomes, and could do so completely reversibly. Moreover, afterwards, at normal temperature, the cells would divide many times over as though nothing had happened. Furthermore, as Ted Salmon showed, the spindle microtubules in living cells could be depolymerized by high hydrostatic pressure and also reversibly so, again exerting pulling or pushing forces on chromosomes and centrosomes as a function of the speed by which the microtubules were falling apart or were being re-polymerized (*Collected Works* A-22, A-25, A-27 Salmon 1975, 1976).

One concern in using polarized light microscopy is the very high intensity of illumination which is required to detect the weak birefringence of spindle fibers and other anisotropic minute regions in living cells (since the specimen is exposed to the intense illumination and most of the light is not removed until it is absorbed by the analyzer). In principle, polarized light microscopy is non-invasive, since one simply records the phase difference induced by the specimen for light polarized in two orthogonal directions (*Collected Works* A-40), and no energy is absorbed. But in reality there could be chromophores inside the cell that do absorb and respond adversely to the strong illumination as seen e.g., in the reaction of the axostyle to visible blue light (*Collected Works* A-33). Also, the red end of the spectrum tends to heat the specimen if the specimen region is exposed to a wide field of illumination (though generally not to point-by-point illumination, since in that case, the point heat is instantly dissipated by conduction through water).

Over the years, therefore, as the source of illumination for polarized light microscopy of living cells, I have used the 546-nm green spectrum emitted by a concentrated-arc mercury lamp, isolated using a high-efficiency narrow-band-pass interference filter combined with an infra-red absorbing filter. The intrinsic intensity, e.g., of the 546-nm line in a 100-Watt concentrated-arc Hg lamp, is sufficiently high for observation and recording the very weak birefringence, and the filtered illumination sufficiently free of the blue and red ends of the spectrum, so that virtually every type of animal and plant cell that I have observed has continued its successive divisions over many hours if placed in an appropriate, temperature-controlled observation chamber (see, e.g., *Collected Works* A-31, A-36, A-41).

Centrifuge Polarizing Microscope (CPM)

As mentioned above, cells can be kept happy and continue their normal physiological or developmental functions, even when placed under some extreme physical conditions. Thus, they can stand unusually high hydrostatic pressures, low (but not high) temperatures, exposure to very high intensity of illumination by a limited wavelength of light, and high centrifugal forces strong enough to stratify organelles within cells or even to generate heavy and light cell fragments.

Taking advantage of this last property, centrifuge microscopes (microscopes capable of observing living cells in density gradient media) of various design have been developed and used since the 1930s. However, none of the many existing types were capable of use for polarization optical observations.

So, in 1997, Bob Knudson and I at MBL tried constructing a prototype Centrifuge Polarizing Microscope (CPM) using a clinical centrifuge motor, to which we attached a specimen-supporting rotor which was made to rotate in synchrony with a flash lamp that acted as the stroboscopic light source. But the axle of the centrifuge rotor was too unsteady to provide a stable support for the specimen chamber, and the flash lamp output was much too weak to generate images of weakly birefringent cellular structures of interest.

I decided to approach Messrs. Iida and Ichie at Hamamatsu Photonic about building a Centrifuge Polarizing Microscope capable of exposing live specimens to up to10,000 × G (gravitational) force, while providing image resolution of no less than 1 μm at the top centrifugal speed. Mr. Hideo Hiruma, Vice President of Hamamatsu Photonics, kindly agreed to support the venture, provided we found a microscope company interested in participating in the project. As it turned out, Keisuke Suzuki from Olympus was then spending a few years carrying out postdoctoral research in our program at MBL (Fig. 5.21b). He convinced engineers from his company to join in the project, and so the collaborative effort between Hamamatsu Photonics, Olympus Optical, and MBL began.

Within a year after the collaboration was agreed upon, we received a finished instrument from Japan (Fig. 5.23a), with the whole functional unit mounted on a vibration isolation bench. The solid marble top of the bench was perforated by an approximately 10-inch diameter hole, penetrated by components of the microscope along one axis and the rotor and its drive along its own separate axis (Fig. 5.23b).

Both axes are arranged vertically, with the microscope mounted on a remote-controlled (CB), motor-driven X, Y, Z mount. Thus, we could observe the region of interest of the spinning specimen as the chosen region came under the objective lens. Above the body tube of the microscope is a specially designed video camera incorporating a built-in interference-fringe-removing plate ahead of the image sensor. The rotor unit is firmly affixed to the bench top.

With the microscope designed jointly by the electronic engineers at Hamamatsu and the optical engineers at Olympus, one could, in fact, generate a maximum speed of 11,700 RPM with the specimen experiencing 11,000 × G force. The finished CPM generated impressively stable images with 1-μm resolution (Fig. 5.24).

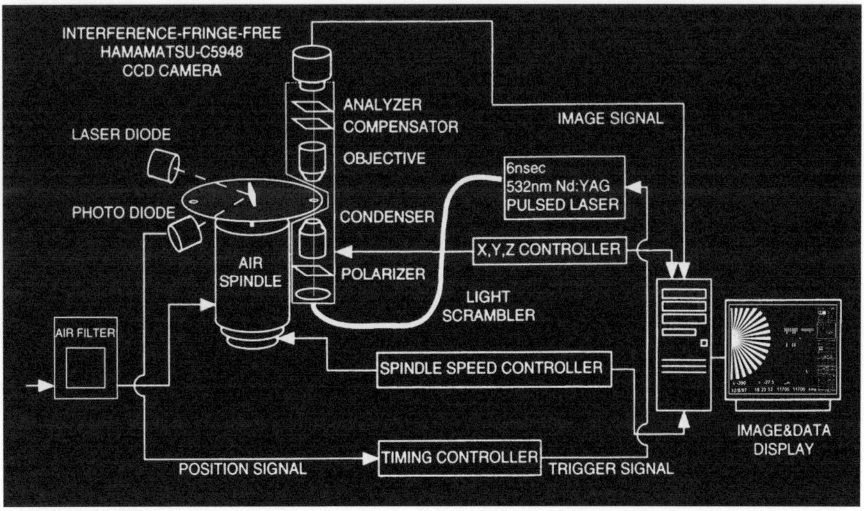

Fig. 5.23a Schematic of Centrifuge Polarizing Microscope (CPM) (Inoué S, Knudson RA, Goda M, Suzuki K, Nagano C, Okada N, Takahashi H, Ichie K, Iida M, Yamanaka K, Centrifuge polarizing microscope. I. Rationale, design, and instrument performance, J Microsc 201 (3): 341–356, 2001. [A 66])

Fig. 5.23b Close-up view of Centrifuge Polarizing Microscope (CPM) showing Rotor (*R*), Photo Diode (*PD*), Diode Laser (*DL*), and Control Box (*CB*) (Inoué S, Knudson RA, Goda M, Suzuki K, Nagano C, Okada N, Takahashi H, Ichie K, Iida M, Yamanaka K, Centrifuge polarizing microscope. I. Rationale, design, and instrument performance, J Microsc 201 (3): 341–356, 2001. [A 66])

The light source chosen by these engineers to obtain such a well synchronized image was a pulsed, argon ion laser delivering 6 nanosecond pulses. For a monochromatic laser pulse of such short duration to be scrambled, the Olympus engineers invented, and chose to use, a special 'golden-hair' array of several hundred thin optical fibers all of different lengths. By travelling such fibers of different length, even the extremely brief monochromatic pulse is scrambled and the mottle, typically imposed on images generated by non-scrambled laser sources, is eliminated.

In practice, however, we found that the brightness of the fibers at the bundled exit of the 'golden-hair' varied from one fiber to another; some of the fibers being definitely bright, while others were darker or completely dark. So in order to illuminate the condenser with a truly uniform bright source, we added an Ellis single-fiber light scrambler to the output of the 'golden hair.' By using this combination, we finally achieved a properly scrambled, steady, uniform background for the image.

The pulses were made to fire exactly when the region of interest in the specimen became coincident with the axis of the objective lens in use. In order to attain this precision of synchrony, a mirror was built into a post above the rotor (R) near its axle (Fig. 5.23b). The mirror reflects the beam from a stationary diode laser (DL) mounted on top of the vibration-isolation bench.

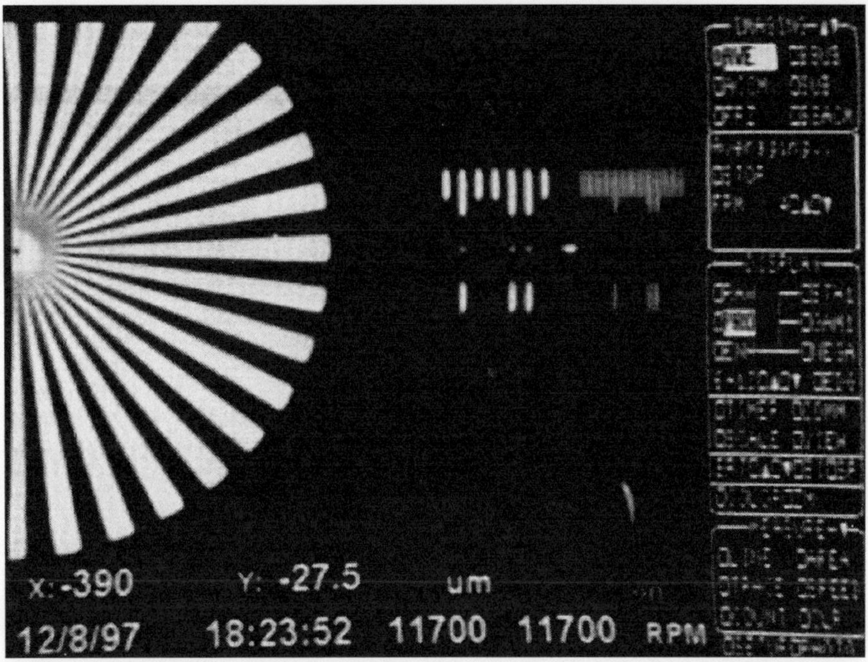

Fig. 5.24 Monitor image of CPM test chart spinning at 11,700 RPM (Inoué S, Knudson RA, Goda M, Suzuki K, Nagano C, Okada N, Takahashi H, Ichie K, Iida M, Yamanaka K, Centrifuge polarizing microscope. I. Rationale, design, and instrument performance, J Microsc 201 (3): 341–356, 2001. [A 66])

A solid-state detector (photo diode, PD), mounted next to the laser, then captures the sweeping beam reflected by the rotor mirror and generates a pulsed electrical signal reflecting the rotating speed and orientation of the rotor. An internal computer compares the frequency and phase of this signal with those generated by the optical encoder built into the rotor. Taking the firing delay of the pulsed source laser into account, the computer further adjusts the signal and provides the final signal required to trigger the source laser exactly on time.

While observation of the specimen using the rotating CPM is carried out using the monochromatic pulsed laser source, preliminary alignment and adjustment of the polarization optical components can only be carried out after the light source is switched to the built-in mercury arc lamp and filters. This automatic switching depends on one of the built in fail-safe devices to prevent accidents to the user or attempts at incompatible use.

For example, when the prism at the base of the body tube (Fig. 5.23b) is switched to the ocular tube for visual observation of the specimen (or of the objective lens back aperture using the Bertrand lens), the laser is switched off and only illumination from the mercury lamp remains. Thus, accidental exposure of the observer's eyes to the intense laser beam is prevented. Similarly, other fail safe devices and

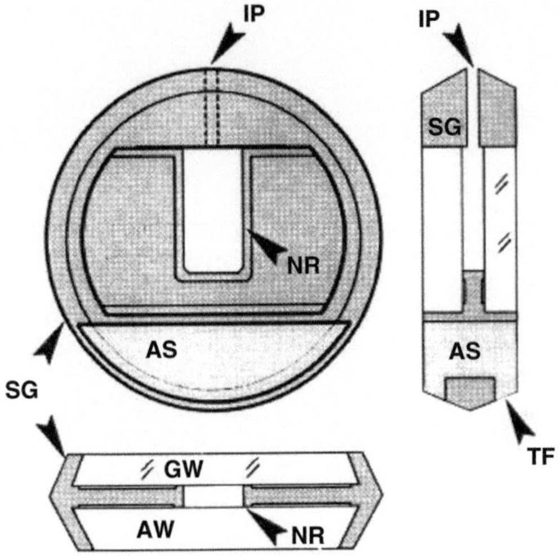

Fig. 5.25 MBL-designed CPM specimen chamber: with input port (*IP*) and leak-proof narrow ridges (NR) built into silicone gasket (*SG*); glass window (*GW*) and acrylic window (*AW*) for automatic birefringence compensation under varying G forces; and acrylic support (*AS*) with tapered flange (*TF*) to match contour of silicon gasket (*Collected Works* A-66) (Inoué S, Knudson RA, Goda M, Suzuki K, Nagano C, Okada N, Takahashi H, Ichie K, Iida M, Yamanaka K, Centrifuge polarizing microscope. I. Rationale, design, and instrument performance, J Microsc 201 (3): 341–356, 2001. [A 66])

circuits are built into the CPM delivered by Hamamatsu and Olympus to prevent misuse or accidental operation of the instrument.

While many parts of the CPM delivered to us functioned well, Bob and I had to re-design the specimen chamber, in addition to improving the uniformity of illumination as already described. This was necessary in order to cut down leakage of cell-bathing fluids, and to eliminate the stress birefringence introduced into the windows, both problems introduced by the centrifugal force on the specimen chamber. As delivered, at several thousand Gs, the chamber started leaking and the stress birefringence of the glass windows of the specimen chamber became several times larger than that of the spindle and completely obscured its image.

In our re-designed specimen chamber (Fig. 5.25), leakage was stopped by using a silicon rubber spacer with built-in small flanges, or ridges (NR), which acts as a seal against the window surfaces. We chose the material for the two windows so that their birefringence was self-compensating. Thus, for the bottom window (AW) we used annealed acrylic (whose **positive** retardance increases with stretch), while for the top window (GW) facing the objective lens, we used tempered glass (whose **negative** retardance increases with stretch).

Using chambers with the self-compensating windows, we could watch and record the density-dependent stratification and fine structural changes in cell organelles. For example, Fig. 5.26a shows an unfertilized sea urchin (*Arbacia*) egg in the CPM concurrently displaying its various stratified organelles (or zones) in compen-

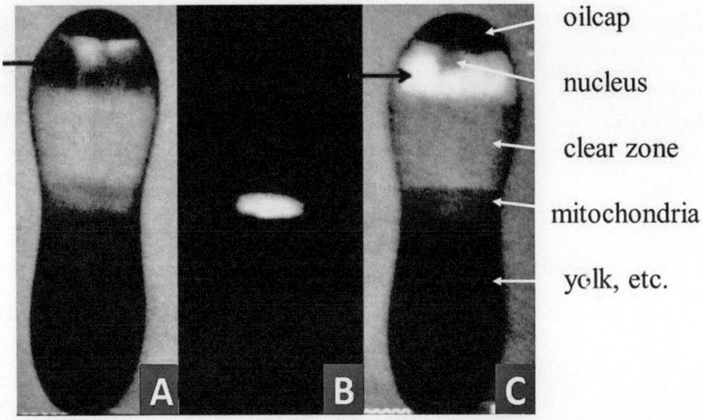

Fig. 5.26a Stratified, unfertilized *Arbacia* egg observed in rotating CPM. Nucleus is seen just below the oil cap and above the birefringent membrane stacks (*yellow arrow* in panels A, C). Fluorescence of Mitotracker-stained mitochondria is seen below clear zone (panels B, C) (*Collected Works* A-67) (Inoué S, Goda M, Knudson RA, Centrifuge polarizing microscope. II. Sample biological applications, J Microsc 201 (3): 357–367, 2001. [A 67])

sated polarization optics, as well as the mitochondrial layer in fluorescence microscopy. To the surprise of some, the pronucleus turns out to be lighter than the endoplasmic reticulum, whose birefringent layers have now become visible, having been centrifugally aligned.

Figure 5.26b shows a *Chaetopterus* oocyte, freed from interfering birefringent granules by centrifuging in a matching density medium. The intact nuclear envelope and the centrifugally-aligned endoplasmic reticulum remain stratified (frames A, B) until the egg is activated. After activation, these structures disintegrate and are replaced by the first division meiotic spindle (frames C, D).

Additional uses of the CPM have included measurement of the climbing ability of the slime mold *Dictyostelium* amoebae in which myosin II was shown to play a critical role in the generation of the cortical contractile force (*Collected Works* A-64). In addition, the CPM has been used to measure adhesive forces between cultured tissue cells; to identify human and Amphiuma red blood cells with varying densities; and to study the adhesiveness of human red cells forming rouleaux and the significance of their reversible shape changes during centrifugal stratification (*Collected Works* A-75).

One additional modification we introduced into the CPM enables the second specimen port to hold a functional specimen chamber. As delivered to us, the second port held a dummy specimen chamber (only for the purpose of counter balancing the rotor). In other words we had been unable to use the dummy chamber to provide a second 'control image' for another specimen experiencing the same centrifugal-force history.

In order to allow the second port to hold a functioning specimen chamber, we added a second timing mirror above the rotor. The second mirror (M-2) faced the opposite direction as the original mirror (M-1), but its face was tilted slightly upward

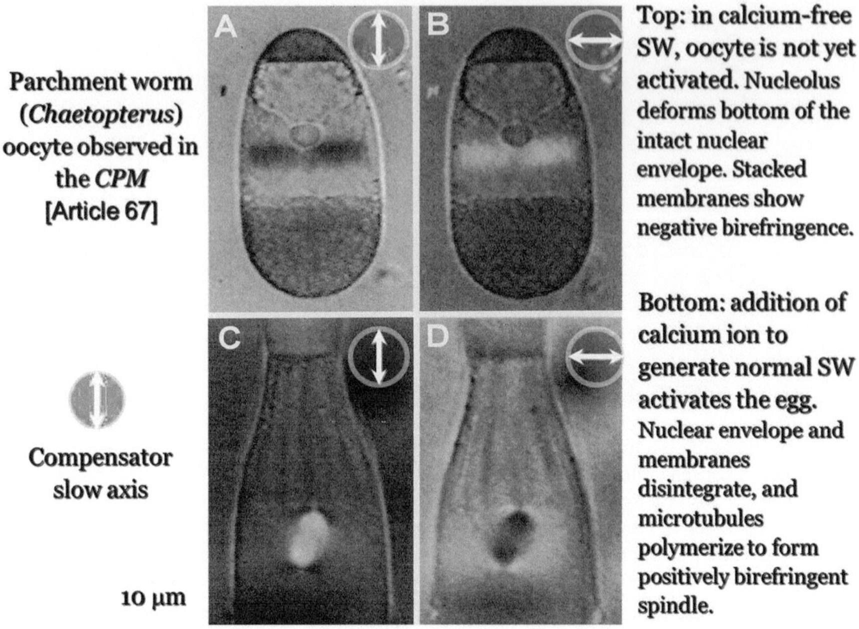

Fig. 5.26b *Chaetopterus* oocyte (Courtesy SI)

from the vertical, in contrast to the original mirror whose face lay in the vertical plane (Fig. 5.27). The timing signal reflected by this second mirror could be diverted, by the beam switching lever (BS), to enter the timing sensor by a switchable set of small mirrors mounted in front of the sensor.

Thus, by selecting the height of the switchable-mirror set (between two preset positions), one could see the synchronized image of the specimen in the original rotor port or the one in the second rotor port. Much to our delight, when the switchable mirror was raised or lowered between the two set positions, one or the other alternate image showed clearly on the video monitor without any delay, and independent of the rotor speed.

Sadly, soon after we had competed this second modification to the CPM, Bob Knudson, who had been flying in seasonally from his new home in Port Townsend, Washington, to help us keep the CPM operating in tiptop shape, succumbed to lung cancer and passed away.

October 2012

Earlier, I had considered closing my recollections here. However, I realized something was missing since I had started this whole narrative referring to the days following the A-bombing of Hiroshima and Nagasaki, but without appropriately emphasizing of their horror.

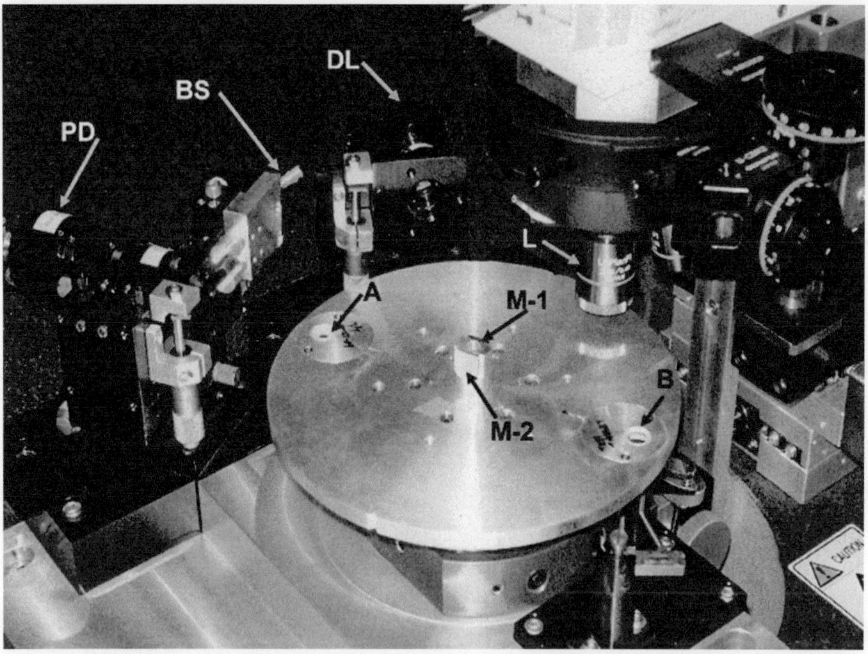

Fig. 5.27 Modified CPM allowing use of two specimen chambers (*A* or *B*). The lever (*BS*) controls the height of the switchable-mirror set that selects which of the two beams, originating from the diode laser (*DL*) and reflected from the timing mirrors (*M*1, *M*2, located on the rotor axis), enters the photodiode (*PD*). The signal from the photodiode, coupled with the signal from the rotor encoder, tells the source laser to fire exactly when the chosen specimen becomes aligned with the center field of the objective lens (*L*) (Courtesy SI)

A couple of nights ago, on October 23, 2012, the Public TV Station WGBH in Boston broadcast a two hour-long program on the 50th anniversary of the Cuban missile crisis. These historic tales reminded us how close we were to a total extermination of all humans, and how the tragedy was averted by the decisive actions of two individuals. One was John F. Kennedy, President of the United States, and the other Commodore Vasili Arkhipov, commander of the four Soviet submarines sent by the Kremlin to attack the U. S. Navy vessels surrounding Cuba.

The story of how Kennedy stood up to Khrushchev, demanding immediate removal of the Soviet nuclear missiles detected through U-2 (?) over flights in Cuba, and his bargaining by removal of the nuclear missiles circling the Soviets, has been told before. On the other hand, I had never heard about the Soviet commander, Vasili Arkhipov, who countermanded the captain of their B-59 submarine and prohibited him from launching its nuclear missile. This incredibly brave action by one officer, taken at the last moment after their subs were forced to surface by the US Navy's depth charges, also played a major role in preventing the world from entering a totally devastating nuclear war.

These touching records shown by WGBH not only revealed the superb statesmanship and cunning of both Kennedy and Khrushchev, facing imminent total disaster for mankind, but also the absolutely critical judgment that these two, as well as the Soviet Commodore, undertook ***as individuals***.

Having lived through these harrowing days of potential nuclear Armageddon, which still hangs over our head, I cannot leave without praying that no single member of our human fellows, or the robots produced by them, will either willfully or by accident ever activate the dreadful fatal switch.

Epilogue

After reminiscing on my upbringing by peace-loving parents in war-torn Japan, and how meeting the Dans injected new hope and purpose into my life despite the War, I spent many words describing how I became wedded to probing some mysteries of living cells using images captured through the polarized light microscope.

But as you may have gathered from the earlier chapters, it wasn't really that I had the foresight to see how powerful the light microscope could become; or to commit a lifetime to developing, or helping to develop, ever more capable light microscopes; or to explore what Nature can tell us about the physical basis of life using such instruments. It was more a culmination of: curiosity, wanting to understand, wanting to share my interests in optics as well as in biology with others; of accidental discovery, response to input from teachers, friends, colleagues and students; and perhaps a bit of competitive spirit that led me through my life (Fig. 5.28). I think this may actually be the way with many scientists.

Fig. 5.28 Shinya at home in 2012 being interviewed by Sara Robberson about his life as a biologist (the interview premiered on the public radio website, Transom.org produced by Sara Robberson, student in Transom's Spring 2012 Story Workshop). Available on the web at http://transom.org/2012/the-cells-mystery/ (Photo: Courtesy C Inoue)

Even though today, still very few biologists take advantage of the polarizing microscope, I believe it is worth emphasizing at this point again how useful and powerful this instrument can be in biology and medicine. It is not only a useful tool for visualizing live tissue and cellular structures that may be otherwise difficult to see, but also for dynamically following the behavior of the molecules that make up even tiny cellular structures. More recent versions of polarized light microscopes are used to follow the dynamic behavior of even single fluorescent molecules inside active living cells.

In fact, I have been blessed with understanding friends and family, excellent co-workers, and even undeserved recognitions. Still rather than dwell further on other events of the last 10 years, this seems to be a reasonable place to stop reminiscing.

As I stop here, I wish to dedicate my memory and share images with the many unforgettable individuals on five continents, who have given me such wonderful opportunities and memorable experiences. I leave with my very best wishes to those of you who have also ventured into this exciting career of probing Nature's ways.

Appendices

Appendix I

The following is the officially provided introduction (p. 115) together with the full text (p. 116–119) of my father's '*KenPaku-Sho*' as reproduced from the publication *Written Statements by Diplomats Who Stood at the Cross Roads before the War*. It has been on exhibit since January, 2007, in the Hall of Historic Records of the Japanese Ministry of Foreign Affairs. To the Japanese reproduction, I (Shinya Inoué) have added my liberal English translation.

10

英・仏を対象とする三国同盟締結への動きに反対する井上庚二郎欧亜局長意見書

ドイツ提案の三国同盟案を、ソ連を主たる対象とする防共協定の延長線上のものとしようとする昭和一四年一月一九日の五相会議における政府方針に対して、あくまで英・仏をその対象に含めようとするドイツはもちろん、我が方陸軍も反対した結果、三月二二日の五相会議で、実質的にはソ連以外の国との戦争に対して日本の武力援助義務を認める妥協案を示した。これに対して当時の井上庚二郎欧亜局長は、妥協案は既決方針の原則の変更であり、独・伊に深くコミットして日本の独自性を奪われる結果となるとの書簡（昭和一四年三月二三日付）を平沼騏一郎首相に送り、妥協案に反対の所信を披瀝した。

115

S. Inoué, *Pathways of a Cell Biologist*, Springer Biographies,
DOI 10.1007/978-981-10-0947-1

今次日独伊三国提携強化の議あり。菲才乏しきを欧亜局長の要職に承け、頭初より之が事務の鞅掌を委せられたるは恩顧洵に浅からざるものあり。日夜黽勉謬なからむことを期し、終始研鑚以て斯案の成立を冀へり。然るに這般在独在伊両帝国使臣に授けられたる訓令に対し両大使より意見具申あり。五省会議は更に議を重ね新しく案を得て今将に訓電を発出せられんとす。

惟ふに白鳥大島両大使が言ふ所は客年来数閲月の長きに亘り、具さに論議を重ね慎重考究せられたる所に係り、縦令予想せられたるが如き両使の稟申あるも、之が為に既定の方針を左右すべきものに非ず。然るに今遽に再議を行ひ、協定事項中蘇連邦の参加せざる場合の英仏両国等を対象とする兵力的援助に付、一度訂立したる方針に一大変革を加へ、曩には之を行ふや否やは一に当時の情況に因るべきものとし、而も現在及近き将来に於て之を行ふ意志なき旨確立したるにも不拘、今次回訓に於ては右兵力的援助を行ふべきことを原則とし、但だ例外として当分の間之を行はざる旨を独伊に通達し、了解を取付け置かむとす。之を換言せば、英仏のみを対象とする兵力的援助を為すべきや否やを、帝国独自の立場より決定せむとしたる方針を変じて、原則的には予め独伊に対し帝国の行動を束縛せむとするものなり。勿論当分の間之が適用なきことに付保留すべしと雖も、右は「当分の間」のことにして、協定全期間を通じて然りとなすことを得ざるべし。若し全期間に亘り此保留を一貫せむとせば、徒らに信義を独伊に失し帝国の威信を堕すものと謂ふべし、或は僅少の援助も援助たるに於ては、即ち一なるを以て、帝国々力の許す以上に援助を行ふこととはならずとの論あるべきも、一旦英仏を対象として兵力を行使せむか、如何に小規模なりと雖も、敵対行為たるは明かなるのみならず、規模範囲自ら拡大し、欲せざるに大規模の敵対行為となるなきを誰か保すべき。殷鑑遠からず今次事変に在り。殊に独外相の言によれば、彼れが帝国に求むる所は欧州に対する派兵に非ざるも、東洋に於て英仏の根拠地を攻撃し、且極東よりする両国通商路の破壊に在り。加之帝国海軍による米国の牽制すら之を欲し居れり。誰か之を小規模且限局せる援助を以て満足するものなりとせむや。

大島大使は独国の強大を知るが故に速きに逮んで之と合縦せむとす。白鳥大使は英国を東亜より駆逐するを以て帝国の使命なりとす。両使が独伊の意向を酌み、我方案を以て非なりとなすは理なきに非ず。何れも熱誠人に勝れ賢良他に劣らざること吾人の疑はざる所なりと雖も、恨むらくは其企図する所直に之を容る可らず。蓋し夫の英たり仏たるもの前世紀白人文明の潮流に棹して東亜の天地に越俎干渉を事とす。素より悪むべしと雖も、蘇狼朔北に吼えて北辺の守り更に急を告ぐるものあり。此重圧を除き、日満両国を磐石の泰きに置かむとせば、姑く隠忍自重して英仏各国を操縦し、有事の日我に左袒せざるまでも敵に与せざるべきを保せざる可らざればなり。若し夫れ我に充実せる力あり後顧の憂なきに至らば、英仏を討ちて東亜の粛正を計るに毫も憚る所なし。今や欧州政局は愈々出でて愈々変転極まる所を知らず。宜しく我に於て能ふ限り行動の自由を確保し置き、一朝有事の際国策の命ずる所に向ひて進退すべし。豈今より之を独伊に約し、異日桎梏の重きを喞つの要あらむや。

斯の独伊の欲する所は協定を天下に公示し、日独伊三国提携の威力を以て各自の意図する政策を実施せむとするに在り。今次我の提案せむとする基本協定は、仮令之を防共協定の延長なりと声明するも、第一条乃至第三条定むる所に依り、世人をして鞏固なる三国枢軸の確立を認識せしむるに足るべきこと、現今の単純なる協定にして既に十分なる効果を挙げ居るに徴し、歴然たりと曰ふべし。即ち帝国原案を其儘に独伊側に提示せむか、之に依り妥結に到達すること必ずしも不可能ならず。要は之を示して接衝大に努むるに在り。然るに今茲に出でず、徒らに前議を飜へして両大使及独伊の意を逆へむとす。惟ふに此膝一び屈すれば復た信ぶ可らず。耿々として事の易きに就くを敢てせば、狂瀾は終に既倒に還すの日なし。君恩の鴻遠なるを浴し大事を托せらるるの眷顧を懐へば、盍ぞ今にして直に両大使に命ずるに我方原案に依る独伊との接洽開始を以てし、両使にして肯かずむば、断然之を召還するの挙に出でざるや。忠誠の士其人に乏しからず、堪能の才豈他に之れなしと言はむや。正を履んで懼れざるは士の道にして又閣下の常に諭へらるる所なり。今此封事を録して特に之を想起せずむば非ず。渺たる小官をも顧みず、敢て此言を為すもの唯々社稷の安泰を希ふの微衷に出づ。而して閣下の赤誠を信ずればなり。庶幾くは明鑒を垂れ、青史に汚点を残さざらむことを。

昭和十四年三月二十三日

井上庚二郎

内閣総理大臣
男爵　平沼騏一郎　閣下

119

English Translation of Father's KenPaku-Sho

This is a liberally translated version by me (Shinya Inoué) of 'Item 10 in: *Written Statements by Diplomats Who Stood at the Cross Roads Before the War*,' which appears in The Hall of Historic Records of the Japanese Ministry of Foreign Affairs, pp. 115–119.

Item 10. Written Opinion by Kojiro Inoué, Director of the Europe and Asia Bureau, Opposing the Move to Accept the Proposal for Japan to Join the Tri-Axis Pact Against England and France.

The plan by the Five Ministers' Council meeting of January 19, 1939: To consider the German proposition "To consider the Triaxis Military Pact as an extension of the Anti-Comintern Treaty" was tabled in view of the opposition by Germany, which wanted to include England and France as potential enemies, as well as opposition by our own Army. In its place, the Five Ministers' Council drafted a compromise plan on March 22: "In principle for Japan to commit military support in case of hostilities against countries other than the Soviet Union." In response to this plan, Kojiro Inoué, then Director of the Europe and Asia Bureau, wrote his objection to Prime Minister Kiichiro Hiranuma on March 23, 1939, pointing out that the compromise plan not only contradicts existing policies but provides a definite commitment to Germany and Italy, resulting in loss of Japan's ability to act independently. [**End of p. 115**]

Body of Text [Starts on p. 116]

Currently there exist deliberation to strengthen the tri-partite cooperation between Japan, Germany, and Italy. Despite my limitations, I have been given the great honor of being entrusted with the important task of directing the Europe and Asia Bureau. While ensuring absolute accuracy of input by day and by night, I have earnestly strived to successfully complete a workable plan. Despite this, our Ambassadors to Germany and Italy have raised a divergent view regarding the directives issued to them. The Five Ministers' Council having further debated this matter is about to wire their new plan to the Ambassadors.

In my opinion, what Ambassadors Shiratori and Ooshima propose relates to what has already been thoroughly discussed and carefully evaluated by us for several months since last year, and even if we receive the anticipated request from the Ambassadors, we must not alter the policy already established. Yet despite the existing policy, there is an imminent plan to completely reverse our position that "In the event that the Soviet Union does not participate, the decision of whether or not our nation provides military assistance against both Great Britain and France will depend on the then prevailing circumstance, and that at present and in the near future, there is no intent by our nation to carry out such a plan." But the wire currently being planned proposes "In principle that Japan agrees to provide military assistance, notifying and receiving agreement from Germany and Italy that only as exception, we would not carry out such a plan for the time being." In other words, this alters the principle that our nation alone decides whether or not to attack England and France, and instead, in principle, leaves it up to Germany and Italy to tie our hands regarding our action. Even though this commitment has the proviso that "For the time being Japan chooses not to use armed forces against Great Britain and France," can such an escape clause be applied indefinitely while the treaty obligation persists? **[Continued to p. 117]** If we attempt to adhere to this escape clause throughout the period the treaty is in effect, we would simply lose trust by Germany and Italy and faith in our Nation. Or, if one insists that even a small amount of aid is still aid, in other words that Japan is still meeting her treaty obligation by providing some military aid within her capability, who can guarantee that the limited engagement will not expand into a major confrontation with the enemy. The signs are clearly obvious. Specifically, even though the Foreign Minister of Germany claims not to desire Japan to send fighting forces to the fronts in Europe, he expects Japan to attack the bases of Great Britain and France in the Orient, and to destroy their shipping routes there. In addition he even desires the Japanese Navy to constrain the United States. Who can claim that such actions define minor, limited actions?

Ambassador Ohshima proposes that we should immediately join hands, being fully aware of Germany's mighty strength. Ambassador Shiratori considers it Japan's obligation to expulse Great Britain from the Orient. It is not surprising that both Ambassadors agree with the views of Germany and Italy and disagree with our own plan. Even though I respect the sincerity and wisdom of both Ambassadors, we must not agree with their points of view. Both Great Britain and France rode the tide of white supremacy in the past century and have interfered with developments in the Far East. This is inexcusable, but we need urgently to protect ourselves from the Soviet wolf howling up North. In order to establish rock solid peace between Manchuria and Japan, by eliminating the major threat from the Soviets, we must for the time being, by exercising patience and prudence, steer Great Britain and France making sure that, even if not as our ally, we do not find them as our enemy. **[Continued to p. 118]** If in the future, we acquire adequate strength and no longer have other concerns, we could then attack Great Britain and France and establish a co-prosperity sphere within the Orient itself. Currently the political situation in Europe is changing rapidly and its outcome is beyond gauging. Today we must retain our freedom to choose and consider our actions, taking into account the international demands facing our nation. We must not agree with the wishes of Germany and Italy today and come to regret the weight of this obligation at a future date.

What Germany and Italy desire is to publically announce the treaty and proceed with their ambitions relying on the mighty power of the cooperation between Japan,

Germany, and Italy. The principle agreement that I plan to propose states that: Even though we declare this as extension of the Anti-Comintern Pact, that it should be clear to the World that the first and third clauses of the Pact show a firm alliance between the three nations, and that this is obvious in view of the fact that the present Pact is already fully effective. It is not out of question that Germany and Italy would accept the present plan of the Japanese Empire, delivered without any modification. What is required is to propose our current plan and engage in strenuous negotiations. In spite of the above, rather than follow this approach, the current (Five Ministers') plan is to alter our pre-existing position and agree with the views of our two Ambassadors and the views of Germany and Italy. But if we yield at this point, it is unlikely that we can ever reverse our position. By choosing the easy way out at this point, the resulting chaos is likely to defeat us in the end. As one entrusted by His Majesty to this great task, I must immediately direct our Ambassadors to commence negotiations with Germany and Italy, and if they disobey, the Ambassadors must be recalled without failure. I do not doubt our Ambassadors' loyalty, nor their wisdom. It is not only the duty of a warrior to proceed along the correct path without fear, but furthermore that is what Your Honor has constantly taught us. I cannot conclude without recalling those words. [**Continued to p. 119**] I thus plead to you despite my insignificant lowly position, only because of my heartfelt concern for the security of our Nation. Furthermore, I trust in Your Honor's sincerity. I pray that with your God-given insight, you will prevent us from leaving a dark spot in our Nation's history.

March 23, 1939 Kojiro Inoué
To The Prime Minister,
Your Excellency, Baron Kiichiro Hiranuma

Appendix II

Address by His Majesty the Emperor

I should like to express my heartfelt congratulations to Dr. Shinya Inoué on being awarded the 19th International Prize for Biology, which is dedicated this year to the field of "Cell Biology."

In an attempt to better understand how cells divide, Dr. Inoué succeeded in making a series of epochal innovations in the development of light microscopy. These advances rendered it possible to directly observe dynamic changes in the supramolecular structure of living cells during cell division. This contributed immensely to advancing research in such fields as cell division, cytoskeleton, and cell motility. Among Dr. Inoué's most remarkable achievements is his use of polarization microscopy to demonstrate that spindles are a real structure within the cell; and that, as a self-assembly system, spindle microtubules are in dynamic equilibrium between assembly and disassembly—a discovery which transformed the prevailing static image of cells to a dynamic one.

Using microscopes that he developed and enhanced himself, Dr. Inoué has achieved a great many research milestones. I have heard, however, that he started his long involvement with microscopes under the severe conditions of Japan's early postwar period. He made use of a discarded rifle pedestal and an old tea can to construct his microscope. When I think of the pains he took back in those days of his youth when Japan was destitute—how he used substitute materials to build a microscope to make his observations—I am deeply impressed by the research efforts that Dr. Inoué continues to exert today, still developing new and ever-improved microscope technologies.

The products of Dr. Inoué's research are widely utilized by researchers around the world, and contribute immensely to the advancement of biological sciences. I sincerely hope that he will enjoy good health and will continue his research activities to achieve improved results.

Appendix III

The 19th International Prize for Biology

ACCEPTANCE ADDRESS BY SHINYA INOUÉ
December 1, 2003

Your Majesties, Ladies and Gentlemen:

It is indeed a great honor today, to be awarded the 2003 International Prize for Biology, this year chosen in the field of Cell Biology. I feel especially privileged to be receiving this prize from your Majesty, who is an accomplished ichthyologist, and celebrating the 60th anniversary of the reign of Emperor Showa, commemorating his life-long devotion to marine biology.

This is doubly a memorable occasion for me personally, since it was shortly after the end of World War II that we welcomed to the Marine Biological Station in Misaki, Emperor Showa together with the Royal Family, including Your Majesty the Crown Prince still in high school, yet already fascinated by the behavior of electric rays.

In Misaki, I was just starting a career in biology under the tutorship of Dan-sensei, yet had the honor of demonstrating to Their Majesties, the twinkling, birefringent spicules in swimming sea urchin larvae, using a polarizing microscope that I had assembled for that occasion.

Again in 1979, when his Majesty Emperor Showa visited the Marine Biological Laboratory in Woods hole, Massachusetts, I had the honor of demonstrating highly magnified images of active hydrozoa. Those exceptionally clear images were seen through the rectified microscope optics that we had developed, jointly with American Optical Company and Nikon.

Starting with the experiences in Misaki, I spent a life time, together with my students and associates, delving into the mysteries of living cells. At the same time, we succeeded in devising new ways of using light waves to explore the living cells' architectural dynamics. I had the privilege of developing new microscopes, and with them, of learning from cells, how to explore the sub-microscopic molecular events that underlay their activities, much of the time at the Marine Biological Laboratory in Woods Hole.

Fortunately our studies on basic cell biology and microscopy have laid the foundation for improved understanding of many essential events of life; including the orderly separation of chromosomes required at every cell division, the transport and positioning of organelles within living cells, the extension and sustenance of nerve cells, and so on. Disorder of these essential cellular events is known to cause many critical diseases as well as genetic and growth anomalies.

Throughout these endeavors, I received indispensable support from many individuals, both in and out of academia, and on both shores of the Pacific. I am especially grateful to my mentors, the late Professors Katsuma and Jean Dan, and Professor Kenneth W. Cooper of Princeton University, who introduced me to the excitement of biological research and to the dignity of the individual. I wish also to thank my many students and colleagues, who have not only learned from but taught me, staff members at the universities and laboratories as well as United States National Science Foundation and National Institutes of Health and the Japan Society for the Promotion of Science, and the many in industry throughout the continents, who have generously provided support, and shared their knowledge and expertise with us.

On this special occasion, I am grateful to be able to share this great honor with my friends and supporters, including my sisters in Japan, Sumiko, Midori, Akiko and Futaba, as well as our children from the States, Heather, Jon, Chris, Steve and Ted, and especially my wife, Sylvia. I also wish to surmise the gratitude and pride that my late parents would have felt, since they both had the honor of serving Emperor and Empress Showa in promoting international cooperation and understanding.

Today, when science has unlocked the atom, probed the edges of the universe, and even decoded the human genome; I too cannot but humbly wish for wiser understanding, and peace, among all men, and greater harmony of Mankind with Nature.

Again, I wish to thank Your Majesties and guests for your kind words and exceptional honor today, and the Selection Committee and my supporters for their sustaining encouragement.

Thank you very much.

Appendix IV

Curriculum Vitae

Dr. Shinya Inoué
Date of Birth: January 5, 1921
Nationality: USA
Position: Distinguished Scientist, Marine Biological Laboratory
Address: Marine Biological Laboratory
7 MBL Street, Woods Hole, MA 02543, USA

Education and Career

1944	B.S. (Zoology) from Tokyo University, Japan
1951	Ph.D. (Biology) from Princeton University, USA
1951–1953	Instructor in Anatomy, University of Washington, Seattle, WA, USA
1953–1954	Assistant Professor of Biology, Tokyo Metropolitan University, Tokyo, Japan
1954–1959	Research Associate to Associate Professor of Biology, Instructor in Optics, University of Rochester, Rochester, NY, USA
1959–1966	Professor and Chairman, Department of Cytology, Dartmouth Medical School, Hanover, NH, USA
1966–1982	Professor of Biology and Director, Program in Biophysical Cytology, University of Pennsylvania, Philadelphia, PA, USA
1979–1987	Instructor-in-Chief, Analytical and Quantitative Light Microscopy, Marine Biological Laboratory, Woods Hole, MA, USA
1986-present	Distinguished Scientist, Cellular Dynamics Program, Marine Biological Laboratory, Woods Hole, MA, USA

Awards and Distinctions

1947, 1953, 1975	Lecture and Demonstrations to Emperor Showa
1955–1958	Scholar in Cancer Research, American Cancer Society
1971–1972	Fellow, John Simon Guggenheim Foundation
1971	Fellow, American Academy of Arts and Sciences
1982–1996	MERIT Award, National Institutes of Health
1986	MBL Distinguished Scientist
1988	Honorary Fellow, Royal Microscopical Society
1992	E. B. Wilson Award, American Society for Cell Biology
1993	Member, National Academy of Sciences
1995	Distinguished Scientist Award, Microscopy Society of America
1997	Ernst Abbe Award, New York Microscopical Society
2003	International Prize for Biology, Japan Society for the Promotion of Science

2010	Order of the Sacred Treasure Gold Rays with Neck Ribbon Award, Government of Japan
2011	Shinya Inoué Symposium on Listening to Life through the Microscope, Marine Biological Laboratory, Woods Hole, MA. USA

Appendix V

A. PUBLICATIONS BY SHINYA INOUÉ APPEARING IN Collected Works of Shinya Inoué: Microscopes, Living Cells, and Dynamic Molecules (including appended DVD). *World Scientific Publishing Co., Singapore, 2008. [Bracketed numbers refer to articles in* Collected Works.*]*

Inoué, S. (1943). Technical news. A device for altering the distance between slide and coverslip at will. *Botany and Zoology Theoretical and Applied* **11** (8), 669. **[A-1]**

Inoué, S. (1944). Japanese Patent #166,528: Stereoscopic apparatus. **[A-2]**

Inoué, S. (1947). Birefringence vs. length and contraction of single muscle fiber. Unpublished. **[A-3]**

Inoué, S. (1951). Coverslip thickness gauge. Unpublished. **[A-4]**

Inoué, S., and K. Dan (1951). Birefringence of the dividing cell. *J. Morphol.* **89** (3), 423–456. **[A-5]**

Inoué, S. (1951). Introduction to doctoral thesis. Revised compilation, 2006. **[A-6]**

Inoué, S. (1951). Thesis Part I. Introduction and description of Shinya Scope-2. Unpublished **[A-7]**

Inoué, S. (1952). Studies on depolarization of light at microscope lens surfaces. I. The origin of stray light by rotation at the lens surfaces[1]. *Exp. Cell Res.* **3** (1), 199–208. (Thesis Part II.) **[A-8]**

Inoué, S. (1951). A method for measuring small retardations of structures in living cells. *Exp. Cell Res.* **2** (3), 513–517. (Thesis Part III.) **[A-9]**

Inoué, S. (1953). Polarization optical studies of the mitotic spindle. I. The demonstration of spindle fibers in living cells. *Chromosoma* **5**, 487–500. (Thesis Part IV.) **[A-10]**

Inoué, S. (1952). Effect of colchicine on the microscopic and submicroscopic structure of the mitotic spindle. *Exp. Cell Res. Suppl.* **2**, 305-318. (Thesis Part V.) **[A-11]**

Inoué, S. (1951). Thesis Part VI. Discussion: Sub-microscopic structure of living spindle. Unpublished. **[A-12]**

Inoué, S. (1952). Effect of temperature on the birefringence of the mitotic spindle. *Biol. Bull.* **103** (2), 316. **[A-13]**

Inoué, S., and W. Lewis Hyde (1957). Studies on depolarization of light at microscope lens surfaces. II. The simultaneous realization of high resolution and high sensitivity with the polarizing microscope. *J. Biophys. Biochem. Cytol.* **3** (6), 831-838. **[A-14]**

Inoué, S., and H. Kubota (1958). Diffraction anomaly in polarizing microscopes. *Nature* **182**, 1725–1726. **[A-15]**

Inoué, S. (1959). Motility of cilia and the mechanism of mitosis. *Rev. Mod. Phys.* **31** (2), 402–408 and *in* J.L. Oncley, ed., <u>In Biophysical Science—A Study Program</u>. John Wiley & Sons, NY, pp. 402–408. **[A-16]**

Kubota, H., and S. Inoué (1959). Diffraction images in the polarizing microscope. *J. Opt. Soc. Amer.* **49** (2), 191–198. **[A-17]**

Inoué, S. (1961). Polarizing microscope: Design for maximum sensitivity. *In* G.L. Clark, ed., The Encyclopedia of Microscopy. Reinhold, NY, pp. 480–485. **[A-18]**

Inoué, S., and A. Bajer (1961). Birefringence in endosperm mitosis. *Chromosoma* **12**, 48–63. **[A-19]**

Inoué, S., H. Sato, and R.W. Tucker (1963). Heavy water enhancement of mitotic spindle birefringence. *Biol. Bull.* **125** (2), 380–381. **[A-20]**

Tucker, R.W. and S. Inoué (1963). Rapid exchange of D_2O and H_2O in sea urchin eggs. *Biol. Bull.* **125** (2), 395. **[A-21]**

Inoué, S. (1964). Organization and function of the mitotic spindle. *In* R.D. Allen and N. Kamiya, eds., Primitive Motile Systems in Cell Biology. Academic Press, New York, NY, pp. 549–598. **[A-22]**

Inoué, S., H. Sato, and M. Ascher (1965). Counteraction of colcemid and heavy water on the organization of the mitotic spindle. *Biol. Bull.* **129** (2), 409–410. **[A-23]**

Inoué, S., and H. Sato (1966). Deoxyribonucleic acid arrangement in living sperm. *In* T. Hayashi and A. Szent-Györgyi, eds., Molecular Architecture in Cell Physiology. Prentice Hall, Englewood Cliffs, NJ, pp. 209–248. **[A-24]**

Inoué, S., and H. Sato (1967). Cell motility by labile association of molecules. The nature of mitotic spindle fibers and their role in chromosome movement. *J. Gen. Physiol.* **50** (6), 259–292. **[A-25]**

Aronson, J., and S. Inoué (1970). Reversal by light of the action of N-methyl N-desacetyl colchicine on mitosis. *J. Cell Biol.* **45** (2), 470–477. **[A-26]**

Inoué, S., J. Fuseler, E.D. Salmon, and G.W. Ellis (1975). Functional organization of mitotic microtubules: Physical chemistry of the in vivo equilibrium system. *Biophys. J.* **15**, 725–744. **[A-27]**

Sato, H., G.W. Ellis, and S. Inoué (1975). Microtubular origin of mitotic spindle form birefringence: Demonstration of the applicability of Wiener's equation. *J. Cell Biol.* **67** (3), 501–517. **[A-28]**

Okazaki, K., and S. Inoué (1976). Crystal property of the larval sea urchin spicule. *Dev. Growth Differ.* **18** (4), 413–434. **[A-29]**

Ritter, H., S. Inoué, and D. Kubai (1978). Mitosis in *Barbulanympha*. I. spindle structure, formation and kinetochore engagement. *J. Cell Biol.* **77** (3), 638–654. **[A-30]**

Inoué, S., and H. Ritter (1978). Mitosis in *Barbulanympha*. II. Dynamics of a two-stage anaphase, nuclear morphogenesis, and cytokinesis. *J. Cell Biol.* **77** (3), 655–684. **[A-31]**

Gordon, G.W., and S. Inoué (1979). Unexpected increase in poleward velocities of mitotic chromosomes after UV irradiation of their kinetochore fibers. *J. Cell Biol.* **83**, 376a. **[A–32]**

Langford, G.M., and S. Inoué (1979). Motility of the microtubular axostyle in *Pyrsonympha*. *J. Cell Biol.* **80** (3), 521–538. **[A-33]**

Inoué, S. (1981). Video image processing greatly enhances contrast, quality, and speed in polarization based microscopy. *J. Cell Biol.* **89** (2), 346–356. **[A-34]**

Inoué, S. (1981). Cell division and the mitotic spindle. *J. Cell Biol.* special issue, "Discovery in Cell Biology," **91** (3), 131 s–147 s. **[A-35]**

Inoué, S., B. Potrebic, C.R. Brown, and D.A. Lutz (1982). Gradient of cleavage inhibition induced by limited diffusion of oxygen. *J. Cell Biol.* **95** (2), 136a. **[A-36]**

Inoué, S., and L.G. Tilney (1982). The acrosomal reaction of *Thyone* sperm. I. Changes in the sperm head visualized by high resolution video microscopy. *J. Cell Biol.* **93** (3) 812–819. **[A-37]**

Tilney, L.G., and S. Inoué (1982). The acrosomal reaction of *Thyone* sperm. II. The kinetics and possible mechanism of acrosomal process elongation. *J. Cell Biol.* **93** (3) 820–827. **[A-38]**

Tilney, L.G., and S. Inoué (1985). Acrosomal reaction of the *Thyone* sperm. III. The relationship between actin assembly and water influx during the extension of the acrosomal process. *J. Cell Biol.* **100**, 1273–1283. **[A-39]**

Inoué, S. (1986). Video Microscopy. Plenum Press, New York. **[Appendix III appears as A-40]**

Lutz, D.A., and S. Inoué (1986). Techniques for observing living gametes and embryos. *In* T. Schroeder, ed., Echinoderm Gametes and Embryos. *Methods Cell Biol.*, Vol. 27. Academic Press, New York, NY, pp. 89–110. **[A-41]**

Inoué, S., and T.D. Inoué (1986). Computer-aided stereoscopic video reconstruction and serial display from high-resolution light-microscope optical sections. *In* Somlyo, ed., Recent Advances in Electron and Light Optical Imaging in Biology and Medicine. *Ann. NY Acad. Sci.*, Vol. 483, pp. 392–404. **[A-42]**

Dan, K., and S. Inoué (1987). Studies of unequal cleavage in molluscs. II. Asymmetric nature of the two asters. *Int. J. Invertebr. Repr. Dev.* **11**, 335–354. **[A-43]**

Inoué, S., R.A. Knudson, and T.D. Inoué (1987). High-N.A. microscope with automatic correction for spherical aberration. *J. Cell Biol.* **105**, 226a. **[A-44]**

Lutz, D.A., Y. Hamaguchi, and S. Inoué (1988). Micromanipulation studies of the asymmetric positioning of the maturation spindle in *Chaetopterus* sp. oocytes: I. Anchorage of the spindle to the cortex and migration of a displaced spindle. *Cell Motil. Cytoskeleton* **11**, 83–96. **[A-45]**

Inoué, S. (1988). The living spindle. *In* K. Dan Festschrift: Advances in Cell Division Research. *Zool. Sci.* **5**, 529–538. **[A-46]**

Inoué, S. (1989). Imaging of unresolved objects, superresolution, and precision of distance measurement, with video microscopy. *In* Taylor and Wang, eds., Fluorescence Microscopy of Living Cells in Culture: Quantitative Fluorescence Microscopy: Imaging and Spectroscopy. *Methods Cell Biol.*, Vol. 30. Academic Press, New York, NY, pp. 85–112. **[A-47]**

Walker, R.A., S. Inoué, and E.D. Salmon (1989). Asymmetric behavior of severed microtubule ends after ultraviolet-microbeam irradiation of individual microtubules in vitro. *J. Cell Biol.* **108**, 931–937. **[A-48]**

Sardet, C., J. Speksnijder, S. Inoué, and L. Jaffe (1989). Fertilization and ooplasmic movements in the ascidian egg. *Development* **105**, 237–249. **[A-49]**

Oldenbourg, R., and S. Inoué (1989). Analysis of edge birefringence observed near refractive index steps in myofibrils and KCL crystals using high resolution polarized light microscopy and spatial Fourier filtering. *Biol. Bull.* **177** (2), 318. **[A-50]**

Inoué, S. (1990). Dynamics of mitosis and cleavage. *In* G.W. Conrad and T.E. Schroeder, eds., Cytokinesis: Mechanisms of Furrow Formation During Cell Division. *Ann. NY Acad. Sci.*, Vol. 582, pp. 1–14. **[A-51]**

Febvre, J., C. Febvre-Chevalier, R. Knudson, T. Takeshita, and S. Inoué (1993). *In vivo* visualization of intercalary microtubule breakdown by polarizing microscopy and very high speed video. Abstract of invited paper presented at IX Intl. Congress of Protozoology, Berlin, Germany, July 25–31, 1993. **[A-52]**

Oldenbourg, R., H. Terada, R. Tiberio, and S. Inoué (1993). Image sharpness and contrast transfer in coherent confocal microscopy. *J. Microsc.* **172** (1), 31–39. **[A-53]**

Inoué, S. (1994). Ultra-thin optical sectioning and dynamic volume investigation with conventional light microscopy. *In* J.K. Stevens, L.R. Mills, and J.E. Trogadis, eds., Three-Dimensional Confocal Microscopy: Volume Investigation of Biological Specimens. Academic Press, pp. 397–419. **[A-54]**

Inoué, S. (1994). Recollections Kayo Okazaki. *Dev. Growth Differ.* **36** (4), 342. **[A-55]**

Inoué, S. (1994). A Tribute to Katsuma Dan. *Biol. Bull.* **187** (2), 127–128. **[A-56]**

Inoué, S. (1995). Foundations of confocal scanned imaging in light microscopy. *In* J. Pawley, ed., Handbook of Biological Confocal Microscopy, 2nd Edition, Chapter 1. Plenum Press, New York, pp. 1–17. **[A-57]**

Inoué, S., & Salmon, E. D. (1995). Force generation by microtubule assembly/disassembly in mitosis and related movements. *Mol. Biol. Cell* **6**, 1619–1640. **[A-58]**

Oldenbourg, R., S. Inoué, R. Tiberio, A. Stemmer, G. Mei, and M. Skvarla (1996). Standard test targets for high-resolution light microscopy. *In* H. Hoch, L. Jelinski, and H. Craighead, eds., Nanofabrication and Biosystems: Integrating Materials Science, Engineering, and Biology, Cambridge University Press, New York, NY, pp. 123–138. **[A-59]**

Fukui, Y., and S. Inoué (1997). Amoeboid movement anchored by eupodia, new actin-rich knobby feet in *Dictyostelium*. *Cell Motil. Cytoskeleton* **36** (4), 339–354. **[A-60]**

Inoué, S., P. Tran, and M. H. Burgos (1997). Photodynamic effect of 488-nm light on Eosin-B-stained *Spisula* sperm. *Biol. Bull.* **193** (2), 225–226. **[A-61]**

Inoué, S., and R. Oldenbourg (1998). Microtubule dynamics in mitotic spindle displayed by polarized light microscopy. *Mol. Biol. Cell* **9** (7), 1603–1607 (video essay). **[A-62]**

Inoué, S. (1999). Windows to dynamic fine structures, then and now. *FASEB J.* **13** (9002), S185–S190. **[A-63]**

Fukui, Y., T.Q.P. Uyeda, C. Kitayama, and S. Inoué (2000). How well can an amoeba climb? *Proc. Natl. Acad. Sci. USA* **97** (18), 10020–10025. **[A-64]**

Burgos, M.H., M. Goda, and S. Inoué (2000). Fertilization-induced changes in the fine structure of stratified *Arbacia* eggs. II. Observations with electron microscopy. *Biol. Bull.* **199** (2), 213–214. **[A-65]**

Inoué, S., R.A. Knudson, M. Goda, K. Suzuki, C. Nagano, N. Okada, H. Takahashi, K. Ichie, M. Iida, and K. Yamanaka (2001). Centrifuge polarizing microscope. I. Rationale, design, and instrument performance. *J. Microsc.* **201** (3), 341–356. **[A-66]**

Inoué, S., M. Goda, and R.A. Knudson (2001). Centrifuge polarizing microscope. II. Sample biological applications. *J. Microsc.* **201** (3), 357–367. **[A-67]**

Knudson, R.A., S. Inoué, and M. Goda (2001). Centrifuge polarizing microscope with dual specimen chambers and injection ports. *Biol. Bull.* **201** (2), 234. **[A-68]**

Shribak, M., S. Inoué, and R. Oldenbourg (2002). Polarization aberrations caused by differential transmission and phase shift in high NA lenses: theory, measurement and rectification. *Opt. Eng.* **41** (5), 943–954. **[A-69]**

Inoué, S., O. Shimomura, M. Goda, M. Shribak, and P.T. Tran (2002). Fluorescence polarization of green fluorescent protein (GFP). *Proc. Natl. Acad. Sci. USA* **99** (7), 4272–4277. **[A-70]**

Inoué, S. (2002). Polarization microscopy. *In* J. Lippincott-Schwartz, ed., Current Protocols in Cell Biology, Vol. 1, Suppl. 13. John Wiley & Sons, pp. 4.9.1–4.9.27. **[A-71]**

Inoué, S., and T. Inoué (2002). Direct-view high-speed confocal scanner—the CSU-10. *In* B. Matsumoto, ed., Cell Biological Applications of Confocal Microscopy, 2nd Edition. Academic Press, pp. 87–123. **[A-72]**

Record of the 2003 International Prize for Biology. **[Appears as Appendices II & III in this volume and A-73]**

Shribak, M., and S. Inoué (2006). Orientation-independent differential interference contrast microscopy. *App. Opt.* **45** (3), 460–469. **[A-74]**

Hoffman, J.F., and S. Inoué (2006). Directly observed reversible shape changes and hemoglobin stratification during centrifugation of human and *Amphiuma* red blood cells. *Proc. Natl. Acad. Sci. USA* **103** (8), 2971–2976. **[A-75]**

B. Other Primary Publications by Shinya Inoué by Year

Inoué, S. (1949). Studies of the *Nereis* egg jelly with the polarization microscope. *Biol. Bull.* **97** (2), 258–259.

Inoué, S. (1959). Problems in polarization microscopy. *J. Opt. Soc. Amer.* **49** (5), 508 (Abstract #FC53).

Inoué, S., and C. J. Koester (1959). Optimum half-shade angle in polarizing instruments. *J. Opt. Soc. Amer.* **49** (6), 556–559.

Hyde, W.L., and S. Inoué (1960). U.S. Patent #2,936,673: Polarizing optical systems.

Inoué, S. (1960). "Birefringence in Dividing Cells." Time-lapse motion picture. Geo. W. Colburn Laboratory, Inc., Chicago, IL.

Inoué, S., and H. Sato (1962). Arrangement of DNA in living sperm: A biophysical analysis. *Science* **136**, 1122–1124.

Reynolds, G.T., R.D. Allen, R. Eckert, J.W. Hastings, and S. Inoué (1963). Application of an image intensifier tube to the microscopic observation of bioluminescent cells and visualization of weak radioactive source distributions. *Biol. Bull.* **125** (2), 389–390.

Inoué, S. (1965). Studies of cell division with an improved polarizing microscope. *In* Von Hippel, ed., The Molecular Designing of Materials and Devices. MIT Press, Cambridge, MA, pp. 211–214.

Inoué, S., and H. Sato (1966). Arrangement of DNA molecules in the sperm nucleus: An optical approach to the analysis of biological structure. *In* Japanese Biophysical Society, ed., Biophysical Science Series 3, Progress in Genetics III. Yoshioka, Kyoto, pp. 151–220 (*in Japanese*).

Stephens, R.E., S. Inoué, and J.I. Clark (1966). Sulfhydryl balance in mitosis: The effect of mercaptoethanol on spindle birefringence. *Biol. Bull.* **131** (2), 409.

Inoué, S. (1969). The physics of structural organization in living cells. *In* S. Devons, ed., Biology and the Physical Sciences. Columbia University Press, New York, NY, pp. 139–171.

Inoué, S., G.W. Ellis, E.D. Salmon, and J.W. Fuseler (1970). Rapid measurement of spindle birefringence during controlled temperature shifts. *J. Cell Biol.* **47**, 95a–96a.

Aronson, J.F., and S. Inoué (1971). Observations on the potential for increasing spindle birefringence during mitosis in *Lytechinus variegatus. J. Gen. Physiol.* **58**, 713.

Langford, G., S. Inoué, and I. Sabran (1973). Analysis of axostyle motility in *Pyrsonympha vertens. J. Cell Biol.* **59**, 185a.

Inoué, S., G.G. Borisy, and D.P. Kiehart (1974). Growth and lability of *Chaetopterus* oocyte mitotic spindles isolated in the presence of porcine brain tubulin. *J. Cell Biol.* **62**, 175–184.

Inoué, S., and H. Ritter (1975). Dynamics of mitotic spindle organization and function. *In* S. Inoué and R.E. Stephens, eds., Molecules and Cell Movement. Raven Press, New York, NY, pp. 3–30.

Inoué, S. (1976). Chromosome movement by reversible assembly of microtubules. *In* R.D. Goldman, T. Pollard, and J. Rosenbaum, eds., Cell Motility, Vol. 3. Cold Spring Harbor Conference on Cell Proliferation, pp. 1317–1328.

Inoué, S., and K. Okazaki (1977). Biocrystals. *Sci. Am.* **236** (4), 82–92.

Inoué, S., and D.P. Kiehart (1978). *In vivo* analysis of mitotic spindle dynamics. *In* Dirksen, Prescott, and Fox, eds., Cell Reproduction, In Honor or Daniel Mazia. Academic Press, New York, NY, pp. 433–444.

Inoué, S., D.P. Kiehart, I. Mabuchi, and G.W. Ellis (1979). Molecular mechanism of mitotic chromosome movement. *In* Pepe, Sanger and Nachmias, eds., Motility in Cell Function, First John M. Marshall Symposium in Cell Biology. Academic Press, New York, NY, pp. 301–311.

Okazaki, K., K. McDonald, and S. Inoué (1980). Sea urchin larval spicule observed with the scanning electron microscope. *In* Omori and Watanable, eds., The Mechanisms of Biomineralization in Animals and Plants. Tokai University Press, Tokyo, pp. 159–168.

Inoué, S., D. Cohen, and G.W. Ellis (1981). High resolution, stereo video microscopy. *Biol. Bull.* **161**, 306.

Inoué, S. (1982a). The role of self-assembly in the generation of biological form. *In* S. Subtelny and P.B. Green, eds., Developmental Order: Its Origin and Regulation, 40th Symposium of the Society for Developmental Biology. Alan R. Liss, New York, NY, pp. 35–76.

Inoué, S. (1982b). Simultaneous video display of fluorescence and polarized light, or DIC, microscope images in real time. *J. Cell Biol.* **95** (2), 461a.

Kiehart, D.P., I. Mabuchi, and S. Inoué (1982). Evidence that myosin does not contribute to force production in chromosome movement. *J. Cell Biol.* **94**, 165–178.

Lutz, D.A., and S. Inoué (1982). Colcemid but not cytochalasin inhibits asymmetric nuclear positioning prior to unequal cell division. *Biol. Bull.* **163** (2), 373–374.

Tanaka, Y., and S. Inoué (1982). Reversible relaxation of cleavage contractile ring by cytochalasins. *J. Cell Biol.* **95** (2), 289a.

Woodruff, R.I., D.A. Lutz, and S. Inoué (1982). Lucifer yellow CH as a non-intrusive, *in vivo* fluorescent probe for physiological studies during early development. *Biol. Bull.* **163**, 379.

Inoué, S., T.D. Inoué, and G.W. Ellis (1983). Visualizing extremely low contrast images by digital enhancement of selected portions of the image gray scale. *Biol. Bull.* **165**, 492.

Hamaguchi, Y., D.A. Lutz, and S. Inoué (1983). Cortical differentiation, asymmetric positioning and attachment of the meiotic spindle in *Chaetopterus pergamentaceous* oocytes. *J. Cell Biol.* **97**, 254a.

Otter, T., D.A. Lutz, J. Eichen, and S. Inoué (1983). Contractile vacuole activity and associated birefringence demonstrated in living *Paramecium caudatum* by video-enhanced DIC and polarized light microscopy. *J. Cell Biol.* **97**, 248a.

Akins, R.E., D.A. Lutz, and S. Inoué (1984). Microtubule assembly and pronuclear movement analyzed by microbeam inactivation of colcemid. *Biol. Bull.* **165**, 501.

Johnson, C.H., S. Inoué, A. Flint, and J.W. Hastings (1984). Compartmentalization of algal bioluminescence: Autofluorescence of bioluminescent particles in the dinoflagellate *Gonyaulax* as studied with image-intensified video microscopy and flow cytometry. *J. Cell Biol.* **100**, 1435–1446.

Inoué, S., A.S. Bajer, J. Molè-Bajer, M. De Brabander, J. De Mey, R. Nuydens, G.W. Ellis, E. Horn, and T.D. Inoué (1985). Microtubules decorated with 5 nm gold visualized by video-enhanced light microscopy. *J. Cell Biol.* **101**, 146a.

Inoué, S., T.D. Inoué, and G.W. Ellis (1985). Rapid, stereoscopic display of microtubule distribution by a video-processed optical sectioning system. *J. Cell Biol.* **101** (2), 146a.

Inoué, S., J. Molè-Bajer, and A.S. Bajer (1985). Three-dimensional distribution of microtubules in *Haemanthus* endosperm cells. *In* De Brabander and De Mey, eds., Microtubules and Microtubule Inhibitors. Elsevier, pp. 269–276.

Ellis, G.W., S. Inoué, and T.D. Inoué (1986). Computer aided light microscopy. *In* P. De Weer and B. Salzberg, eds., Optical Methods in Cell Physiology. John Wiley and Sons, New York, NY, pp. 15–30.

Inoué, S. (1987). Video microscopy of living cells and dynamic molecular assemblies. *Appl. Opt.* **26** (16), 3219–3225.

Inoué, S. (1987). Ultrathin optical-sectioning-tomography achieved with the light microscope. *Biol. Bull.* **173**, 419.

Tilney, L.G., and S. Inoué (1987). Flagellar gyration and midpiece rotation during extension of the acrosomal process of *Thyone* sperm: How and why this occurs. *J. Cell Biol.* **104**, 407–415.

Molè-Bajer, J., A.S. Bajer, and S. Inoué (1988). Three-dimensional localization and redistribution of F-actin in higher plant mitosis and cell plate formation. *Cell Motil. Cytoskeleton* **10**, 217–228.

Cassimeris, L., S. Inoué, and E.D. Salmon (1988). Microtubule dynamics in the chromosomal spindle fiber: Analysis by fluorescence and high-resolution polarization microscopy. *Cell Motil. Cytoskeleton* **10**, 185–196.

Inoué, S. (1988). Progress in video microscopy. *Cell Motil. Cytoskeleton* **10**, 13–17.

Inoué, S. (1988). Manipulating single microtubules. *Protoplasma Suppl.* **2**, 57–62.

Inoué, S., and T. Inoué (1988). Stereoscopy of high resolution video microscope images reconstructed from serial optical sections. *Biol. Bull.* **175**, 316.

Inoué, S., T. Inoué, R.A. Knudson, and R. Oldenbourg (1989). Very high resolution and dynamic stereo images of neurons. *Biol. Bull.* **177** (2), 322.

Silver, R.E., O. Shimomura, and S. Inoué (1989). Detection of intracellular free calcium with aequorin and single photon video light microscopy. *J. Cell Biol.* **109**, 90a.

Inoué, S. (1990). Foundations of confocal scanned imaging in light microscopy. *In* J. Pawley, ed., Handbook of Biological Confocal Microscopy, Chapter 1. Plenum Press, New York, NY, pp. 1–14.

Inoué, S. (1990). Whither video microscopy? Towards 4-D imaging at the highest resolution of the light microscope. *In* B. Herman and K. Jacobson, eds., Optical Microscopy for Biology. Wiley-Liss, New York, NY, pp. 497–511.

Kiehart, D.P., A. Ketchum, P. Young, D. Lutz, M.R. Alfenito, X.-J. Chang, M. Awobuluyi, T.C. Pesacreta, S. Inoué, C.T. Stewart, and T.-L. Chen (1990). Contractile proteins in *Drosophila* development. *In* G.W. Conrad and T.E. Schroeder, eds., Cytokinesis: Mechanisms of Furrow Formation During Cell Division. *Ann. NY Acad. Sci.*, Vol. 582, pp. 233–251.

Sanger, J.M., and J.W. Sanger, video eds. (1990). Video Supplement 2. Experiments on the mitotic spindle viewed in polarized light (S. Inoué, A.S. Bajer, J. Fuseler, & E.D. Salmon). Labile attachment of meiotic spindle to localized site on cell cortex (Y. Hamaguchi, D.A. Lutz, & S. Inoué). Reversible inhibition of pronuclear migration in a sea urchin, *Lytechinus variegatus* (J.F. Aronson & S. Inoué). Birefringence of flagella and axostyle in hypermastigotes (S. Inoué & V. Chen). *Cell Motil. Cytoskeleton* **17**, 356–372.

Fukui, Y., and S. Inoué (1991). Cell division in *Dictyostelium* with special emphasis on actomyosin organization in cytokinesis. *Cell Motil. Cytoskeleton* **18**, 41–54.

Inoué, S. (1993). Porter and the fine architecture of dividing cells. *In* R.B. Barlow, Jr., J.E. Dowling, and G. Weissmann, eds., The Biological Century: Friday Evening Talks at the Marine Biological Laboratory. Harvard University Press, pp. 100–115.

Inoué, S. (1993). Video Microscopia. Inoué's 1986 Video Microscopy translated into Spanish by Mario H. Burgos. Inca Editorial, Mendoza, Argentina.

Inoué, S., and T.D. Inoué (1994). Through-focal and time-lapse stereoscopic imaging of dividing cells and developing embryos in DIC and polarization microscopy. *Biol. Bull.* **187**, 232–233.

Inoué, S., and R. Oldenbourg (1995). Optical instruments. Microscopes. *In* Optical Society of America, ed., Handbook of Optics, 2nd Edition, Vol. 2. McGraw-Hill, Inc., pp. 17.1–17.52.

Inoué, S., and T. Inoué (1996). Digital unsharp masking reveals fine detail in images obtained with new spinning-disk confocal microscope. *Biol. Bull.* **191**, 269–270.

Inoué, S. (1996). Mitotic organization and force generation by assembly/disassembly of microtubules. *Cell Struct. Funct.* **21**, 375–379.

Inoué, S., (1996). To Professor Katsuma Dan (Obituary). *Dev. Growth Differ.* **38**, 450–451.

Inoué, S. (1997). The role of microtubule assembly dynamics in mitotic force generation and functional organization of living cells. *J. Struct. Biol.* **118**, 87–93.

Inoué, S., and K. R. Spring (1997). Video Microscopy–The Fundamentals, 2nd ed. Plenum Press, New York, NY.

Tran, P.T., E.D. Salmon, and S. Inoué (1997). UV cutting of MAPs-bound microtubules. *Biol. Bull.* **193**, 218–219.

Tran, P.T., A. DePina, S. Inoué, R. Oldenbourg, S.J. Segal, and M.H. Burgos (1997). Effect of gossypol on *Spisula* sperm observed with real-time confocal microscopy, polarized light microscopy, and video microscopy. *Biol. Bull.* **193**, 227–228.

Inoué, S., M. Goda, G. Gundersen, and R.A. Knudson (1998). Oocyte maturation and tissue cell adhesion observed with centrifuge polarizing microscope. *Mol. Biol. Cell* **9**, 105a.

Goda, M., S. Inoué, and R.A. Knudson (1998). Oocyte maturation in *Chaetopterus pergamentaceous* observed with centrifuge polarizing microscope. *Biol. Bull.* **195**, 212–214.

Inoué, S., R.A. Knudson, K. Suzuki, N. Okada, H. Takahashi, M. Iida, and K. Yamanaka (1998). Centrifuge polarizing microscope. *Microsc. Microanal.* **4** (Suppl. 2), 36–36 (abstract).

Fukui, Y., S. Engler, S. Inoué, and E.L. De Hostos (1999). Architectural dynamics and gene replacement of coronin suggest its role in cytokinesis. *Cell Motil. Cytoskeleton* **42**, 204–217.

Fukui, Y., E.L. De Hostos, S. Yumura, T. Kitanishi-Yumura, and S. Inoué (1999). Architectural dynamics of F-actin in eupodia suggests their role in invasive locomotion in *Dictyostelium. Exp. Cell Res.* **249**, 33–45.

Fukui, Y., T.Q.P. Uyeda, C. Kitayama, and S. Inoué (1999). Migration forces in *Dictyostelium* measured by centrifuge DIC microscopy. *Biol. Bull.* **197**, 260–262.

Maddox, P., A. Desai, E.D. Salmon, T.J. Mitchison, K. Oogema, T. Kapoor, B. Matsumoto, and S. Inoué (1999). Dynamic confocal imaging of mitochondria in swimming *Tetrahymena* and of microtubule poleward flux in *Xenopus* extract spindles. *Biol. Bull.* **197**, 263–265.

Suzuki, K., and S. Inoué (1999). U.S. Patent #5930033: Slit scan centrifuge microscope.

Inoué, S. (1999). U.S. Patent #5982535: Centrifuge microscope capable of realizing polarized light observation.

Goda, M., M.H. Burgos, and S. Inoué (2000). Fertilization-induced changes in the fine structure of stratified *Arbacia* eggs. I. Observations on live cells with the centrifuge polarizing microscope. *Biol. Bull.* **199** (2), 212–213.

Tran, P.T., L. Marsh, V. Doye, S. Inoué, and F. Chang (2001). A mechanism for nuclear positioning in fission yeast based on microtubule pushing. *J. Cell Biol.* **153**, 397–411.

Inoué, S., and M. Goda (2001). Fluorescence polarization of GFP crystals. *Biol. Bull.* **201** (2), 231–233.

Shribak, M., S. Inoué, and R. Oldenbourg (2002). Rectifiers for suppressing depolarization caused by differential transmission and phase shift in high NA lenses. *In* D. Goldstein, D. Chenault,

W. Egan, and M. Duggin, eds., Polarization Analysis, Measurement, and Remote Sensing IV. *Proceedings of Society of Photo-Optical Instrumentation Engineers* **4481**, 163–174.

Shribak, M., S. Inoué, and R. Oldenbourg (2002). Polarization aberrations caused by differential transmission and phase shift in high NA lenses: theory, measurement and rectification. *Opt. Eng.* **41** (5). 943–954.

Inoué, S. (2003). Exploring living cells and molecular dynamics with polarized light microscopy. *In* P. Török and K. Kao, eds., Optical Imaging and Microscopy, Vol. 87. Springer-Verlag, Berlin Heidelberg, pp. 3–20.

Inoué, S. (2006). In memory of Hidemi Sato. *Biol. Cell.* **98** (10), e1–e2.

Shribak, M., J. LaFountain, D. Biggs, and S. Inoué (2007). Quantitative orientation-independent differential interference contrast (DIC) microscopy coupled with orientation-independent polarization microscopy. *In* C. Lyman et al., eds., Proceedings of Microscopy and Microanalysis 2007. Cambridge University Press, pp. 10–12.

Shribak, M., J. LaFountain, D. Biggs, and S. Inoué (2007). Quantitative orientation-independent differential interference contrast (DIC) microscopy. *In* Imaging, Manipulation, and Analysis of Biomolecules, Cells, and Tissues V, D.L. Farkas et al, Editors, *Proc. SPIE* 6441, p. 64411 L1–6441 L12.

Shribak, M., J. LaFountain, D. Biggs, and S. Inoué (2008). Orientation-independent differential interference contrast microscopy and its combination with an orientation-independent polarization system. *Journal of Biomedical Optics* **13** (1), 014011.

Inoué, S. (2008a). Collected Works of Shinya Inoué. World Scientific Publishing Co., Singapore.

Inoué, S. (2008b). Microtubule Dynamics in Cell Division: Exploring Living Cells with Polarized Light Microscopy. *Ann. Rev. Cell Dev. Biol.* **4**, 1–28.

Goda, M., S. Inoué, and I. Mabuchi (2009). Calyculin-A induces cleavage in a random plane in unfertilized sea urchin eggs. *Biol. Bull.* **216**: 40–44.

C. Other Publications Referenced in This Volume

Allen, R. D., N. S. Allen, and J. L. Travis (1981). Video-enhanced differential interference contrast (AVEC-DIC) microscopy: A method capable of analyzing microtubule-related movement in the reticulopodial network of *Allogromia laticollaris. Cell Motility* **1**, 291–302.

Banga, I. & Szent-Györgyi, A. (1941–42). Preparation and properties of myosin A and B. *In* A. Szent-Györgyi, ed., Studies from the Institute of Medical Chemistry University Szeged, Vol. 1. S. Karger AG, Basel, pp. 5–15.

Bělăr, K. (1926). Der Formwechsel der Protistenkerne. *Ergebn. Fortschr. Zool.* **6**, 1–420.

Borisy, G. G. & Olmsted, J. B. (1972). Nucleated assembly of microtubules in porcine brain extracts. *Science* **177**, 1196–1197.

Cook, H. T., and T. F. Cook (1992). Japan at War: An Oral History. The New Press, New York.

Coue, M., V. A. Lombillo, & J. R. McIntosh (1991). Microtubule depolmerization promotes particle and chromosome movement *in vitro. J. Cell Biol.* **112**, 1165–1175.

Dan, J. C., A. Kitahara, and T. Kohri (1954). Studies on the acrosome. II. Acrosome reaction in starfish spermatozoa. *Biol. Bull.* **107**, 203–218.

Dell, K. R., and R. D. Vale (2004). A tribute to Shinya Inoué and innovation in light microscopy. *J. Cell Biol.* **165** (1), 21–25.

Ellis, G. W. (1966). Holomicrography. Transformation of image during reconstruction *a posteriori. Science* **154**, 1195–1196.

Ellis, G. W. (1979). A fiber-optic phase-randomizer for microscope illumination by a laser. *J. Cell Biol.* **83**, 303A.

Grell, K. G., ed. (1973). Protozoology, 3rd Ed. Springer-Verlag, Berlin, Fig. 418.

Huxley, A. F., and R. Niedergerke (1954). Structural changes in muscle during contraction: Interference microscopy of living muscle fibres. *Nature* **173**, 971–973.

Huxley, H., and J. Hanson (1954). Changes in the cross-striations of muscle during contraction and stretch and their structural interpretation. *Nature* **173**, 973–976.

Hyman, L. H. (1940). The Invertebrates: Protozoa through Ctenophore. McGraw-Hill, New York, Fig. 63C.

Inoue, K. (1991). McArthur's Japanese Constitution: A Linguistic and Cultural Study of its Making. University of Chicago Press, Chicago.

Mazia, D., and K. Dan (1952). The isolation and biochemical characterization of the mitotic apparatus of dividing cells. *Proc. Natl. Acad. Sci. USA* **38** (9), 826–838.

Oldenbourg, R. (1991). Analysis of edge birefringence. *Biophys. J.* **60**, 629–641.

Oldenbourg, R. (1996). A new view on polarization microscopy. *Nature* **381**, 811–812.

Oldenbourg, R., and G. Mei (1995). New polarized light microscope with precision universal compensator. *J. Microscopy* **180** (2), 140–147.

Porter, K. R. (1966). Cytoplasmic microtubules and their function. *In* G. E. W. Wolstenholme and M. O'Connor, eds., Ciba Foundation Symposium on Principles of Biomolecular Organization. Churchill Ltd., London, pp. 308–345.

Powell, C., with J. Persico (1995). My American Journey. Random House, New York.

Salmon, E. D. (1975). Spindle microtubules: Thermodynamics of *in vivo* assembly and role in chromosome movement. *Ann. NY Acad. Sci.* **253**, 383–406.

Salmon, E. D. (1976). Pressure-induced depolymerization of spindle microtubules. IV. Production and regulation of chromosome movement. *In* R. Goldman, T. Pollard, and J. Rosenbaum, eds., Cell Motility, Cold Spring Harbor Conferences of Cell Proliferation, Vol. 3. Cold Spring Harbor Laboratory Press, Cold Spring Harbor, NY, pp. 1329–1342.

Salmon, E. D., and G. W. Ellis (1975). A new miniature hydrostatic pressure chamber for microscopy: Strain-free optical glass windows facilitate phase contrast and polarized light microscopy of living cells. Optional fixture permits simultaneous control of pressure and temperature. *J. Cell Biol.* **65**, 587–602.

Schmidt, W. J. (1937). Die Doppelbrechung von Karyoplasma, Zytoplasma und Metaplasma. Gebrüder Borntraeger, Berlin.

Szent-Györgyi, A. (1945). Studies on muscle. *Acta Physiol. Scandinav.* **9** (Suppl.), 25.

Thornburg, W. (1955). Reflecting objective for microspectroscopy. *J. Opt. Soc. Am.* **45**, 740–743.

Tyndall, J. (1867). Sound: A Course of Lectures Delivered at the Royal Institution of Great Britain. D. Appleton and Company, New York.

Weisenberg, R. C. (1972). Microtubule formation *in vitro* in solutions containing low calcium concentrations. *Science* **177**, 1104–1105.

Wright, F. E. (1911). The Methods of Petrographic-Microscope Research, Their Relative Accuracy and Range of Application. Carnegie Institution of Washington, Washington.

MIX
Papier aus verantwortungsvollen Quellen
Paper from responsible sources
FSC® C105338

If you have any concerns about our products,
you can contact us on
ProductSafety@springernature.com

In case Publisher is established outside the EU,
the EU authorized representative is:
Springer Nature Customer Service Center GmbH
Europaplatz 3, 69115 Heidelberg, Germany

Printed by Libri Plureos GmbH
in Hamburg, Germany